W0079196

Mastering Innovation in China

Joachim Jan Thraen

Mastering Innovation in China

Insights from History on China's
Journey towards Innovation

 Springer Gabler

Joachim Jan Thraen
Zurich, Switzerland

Dissertation University of St. Gallen, Switzerland, 2015

ISBN 978-3-658-14555-2   ISBN 978-3-658-14556-9 (eBook)
DOI 10.1007/978-3-658-14556-9

Library of Congress Control Number: 2016943414

Springer Gabler
© Springer Fachmedien Wiesbaden 2016

Printed on acid-free paper

This Springer Gabler imprint is published by Springer Nature
The registered company is Springer Fachmedien Wiesbaden GmbH

Abstract

In recent years, China's economic growth model has started to undergo a fundamental shift away from an export-led capital investment focus toward domestic consumption, productivity and higher levels of innovation. In this transition, China's government has used a range of policies and other measures to increase indigenous (i.e. China-based) innovation in an effort to turn China into a global leader in R&D and innovation, e.g. by committing about 1.7 trillion US dollars to support a selection of seven "Strategic Emerging Industries" (SEI) - for example renewable energy - that are considered as key sectors that will drive China's future economic growth and prosperity. The share of these industries in China's GDP is supposed to increase to eight percent in 2015 and to 15 percent by 2020.

Despite an increasing number of contributions reflecting the growing significance of China as a hub for global innovation, these have largely failed to take a long-term perspective to assess these ongoing developments as well as the country's overall innovation trajectory. One way to use such an approach is to ask how a *historical perspective* on China's government policies to support indigenous innovation in Strategic Emerging Industries can increase our understanding of relevant theoretical and managerial issues in the current debate.

The present study investigates this question. It examines economic, institutional and sociocultural aspects of innovation systems across different geographies and time periods – from the First and Second Industrial Revolution up to the period following the Second World War – and applies the gained insights to four case studies of Chinese and foreign corporations: Covestro (formerly Bayer Material Science); Daimler-BYD; Haier Group; and Siemens AG.

This book integrates the historical element into the current debate on innovation in China. It shows that due to its size and diversity, China is undergoing a transition towards a global innovation hub at different speeds depending on region and industry. For companies, China's large market, strong manufacturing networks, increasing R&D capabilities and a government strongly supporting innovation provide unique opportunities for new forms of innovation driven by efficiency, rapid commercialization and large volumes. Managers that understand China's innovation trajectory and adjust innovation strategies accordingly will achieve greater success in mastering innovation in China as a foundation for global competitiveness.

Zusammenfassung

In den vergangenen Jahren befindet sich China in einem fundamentalen Transformationsprozess von einem export- und investitionsgestützten Wirtschaftsmodell hin zu einer Volkswirtschaft, die stärker auf Binnennachfrage, höhere Produktivität und Innovationsfähigkeit setzt. In diesem Übergangsprozess setzt die chinesische Regierung eine Reihe von Maßnahmen ein, um Innovation zu fördern und China zu einem global führenden Standort für Forschung, Entwicklung und Innovation zu machen. Zwischen 2010 und 2015 stellt sie Fördermittel in Höhe von 1,7 Billionen US-Dollar bereit, um sieben ausgewählte Strategische Wachstumsindustrien („„Strategic Emerging Industries" oder „SEI") – etwa in Erneuerbaren Energien – zu fördern. Diese werden als Schlüsselindustrien für Chinas langfristiges Wachstum angesehen. Ihr Anteil am Bruttoinlandsprodukt soll sich bis Ende 2015 auf acht Prozent und bis 2020 auf 15 Prozent erhöhen.

Eine steigende Anzahl an Beiträgen behandelt die wachsende Bedeutung Chinas als globaler Innovationsstandort. Bislang fehlt aber eine systematische langzeitliche Perspektive, welche die derzeitigen Entwicklungen adäquat in ihren (historischen) Gesamtzusammenhang einordnet.

Die vorliegende Arbeit geht dieser Frage nach. Sie untersucht wirtschaftliche, institutionelle und soziokulturelle Aspekte von Innovationssystemen über verschiedene Regionen und historische Zeiträume – von der Ersten und Zweiten Industriellen Revolution bis zur Zeit nach dem Zweiten Weltkrieg – und wendet die hieraus gewonnenen Erkenntnisse auf vier Fallstudien von chinesischen und ausländischen Unternehmen an, die in Strategischen Wachstumsindustrien in China Forschung und Entwicklung betreiben: Covestro (bis 31. August 2015 Bayer MaterialScience AG); Daimler-BYD; Haier Group; und Siemens AG.

Das Buch integriert historische Erkenntnisse in die derzeitige Debatte über Innovation in China. Es zeigt auf, dass sich China aufgrund seiner Größe und Diversität je nach Region und Industrie mit unterschiedlicher Geschwindigkeit zu einem globalen Innovationsstandort entwickelt. Chinas großer Markt, starke Industrienetzwerke, wachsende Fähigkeiten in Forschung und Entwicklung und eine Regierung, die Innovation stark fördert, bieten Unternehmen Möglichkeiten für neue Formen der Innovation, die auf Effizienz, schnelle und kundenzentrierte Vermarktung in einem großen Markt beruhen. Manager, die Chinas Innovationspfad verstehen und ihre Strategien entsprechend anpassen, werden mehr Innovation in China erzielen können als Grundlage für globale Wettbewerbsfähigkeit.

Table of contents

Abbreviations

Abbreviation	Meaning
BMS	Bayer MaterialScience
BU	Business Unit
BYD	*Build Your Dreams* (automotive brand name)
CAS	Coatings, Adhesives, Specialities (Bayer MaterialScience)
CT	Corporate Technology (department within Siemens AG)
CT	Computer Tomography
ECG	Electro cardiography
EU	European Union
FDI	Foreign Direct Investment
GDP	Gross Domestic Product
GERD	Gross domestic expenditures on research & development (R&D)
IPO	Initial Public Offering
IPR	Intellectual Property Right
IT	Internet Technology
KPI	Key Performance Indicator
L&MR	Liverpool and Manchester Railway
MIIT	Ministry of Industry and Information Technology (China)
MITI	Ministry for Trade and Industry (Japan)
MNC	Multinational Corporation
MNE	Multinational Enterprise
MOFCOM	Ministry of Commerce (China)
MOST	Ministry of Science and Technology (China)
NDRC	National Development and Reform Commission (China)
NEV	New-Energy Vehicle
NIH	National Institutes of Health (USA)
NSI	National System of Innovation
OECD	Organisation for Economic Co-operation and Development
PCS	Polycarbonate (Bayer MaterialScience)
PPP	Purchasing Power Parity
PRC	People's Republic of China

Abbreviation	Meaning
PRDC	Polymer Research and Development Center (Bayer MaterialScience)
PUR	Polyurethanes (Bayer MaterialScience)
R&D	Research & Development
SEI	Strategic Emerging Industries
SIPB	State-Created Intellectual Property-Related Barrier
SIPM	State-Created Intellectual Property-Related Measure
SOE	State-Owned Enterprise
STI	Science, Technology and Innovation
TRIPS	Agreement on Trade-Related Aspects of Intellectual Property Rights
US/USA	United States/United States of America
WTO	World Trade Organization

Tables

Figures

1 Introduction

> "He who innovates will have for his
> Enemies all those who are well off under the
> Existing order of things, and only
> Lukewarm supporters in those who might be
> Better off under the new."
>
> (Machiavelli, *The Prince*, 1513)

In recent years, a number of large emerging economies have experienced rapid economic growth, emerging as important pillars of the global economy. At the same time, some of these countries are becoming important centers of global research and development (R&D) and innovation activities. Evidence such as increasing numbers of patent filings in emerging markets suggests that the geographic locus of innovation is gradually shifting from the traditional, advanced economy centers of innovation towards less developed economies (Bruche, 2009; von Zedtwitz, 2004). As a consequence, multinational companies (MNCs) increasingly locate their R&D activities in developing countries, with proximity to the most dynamic consumer markets. However, as previous studies have shown, the institutional, economic, political, and social context characteristic of many emerging markets have important repercussions for companies engaging in innovation-related activities in those markets (Boisot & Child, 1996; Bruton, Ahlstrom, & Puky, 2009; Bruton & Ahlstrom, 2003; Child & Tse, 2001).

Due to its particular economic and institutional structure and its large and dynamic market, one emerging economy that is of particular interest is China. Featuring the world's largest population, 1.3 billion people, as well as a large and growing middle class of consumers, China has emerged as a significant center for global R&D and innovation. From 1997 to 2011, the number of foreign R&D centers in China increased from 24 in 1997 to around 1,500 in 2011(Yip & McKern, 2014). As manufacturing wages have been rising rapidly in Mainland China over the past decades, from the mid-2000s, China's central government has declared its ambition to move the domestic economy away from its traditional focus on low-wage, export-based manufacturing and towards more innovation and technology-intensive growth industries. For instance, this has been reflected in its latest Five-Year Plan (China Central Government, 2011). The resulting policies and regulations are aimed at narrowing China's technolo-

gy gap with world leaders, in an effort to move beyond the middle-income trap
and to build up long-term sustainable competitiveness.

Central to China's effort to become a global leader in innovation and R&D
are government policies to increase the domestically produced level of innovation, known as "indigenous innovation", and referred to in Chinese as 自主创新
or zìzhǔ chuàngxīn (McGregor, 2010). As an important part of this ambition, the
Chinese central government has selected a number of highly-innovative industries that it hopes will propel China's economic development to a new level,
while also addressing increasing socio-economic and environmental challenges
at home, which may be exacerbated if China does not maintain sustainable
growth rates in the mid- and long-term (USCBC, 2013). These seven industries
are commonly referred to as "Strategic Emerging Industries" (SEI) and include
the following: energy efficient and environmental technologies; next generation
information technology (IT); biotechnology; high-end equipment manufacturing;
new energy; new materials; as well as new-energy vehicles (NEV) such as battery or fuel cell powered vehicles (China Central Government, 2011). Previous
studies have started to investigate the impact of innovation-related government
policies to support domestic innovation, especially in SEI, on foreign companies'
commercial opportunities in China, with a particular focus on US businesses
(USCBC, 2013).

The existing literature has provided some knowledge about how the political and institutional environment of China affects organizations with respect to
innovation (Keupp, Palmié, & Gassmann, 2012), and in particular how European
businesses as well as Chinese companies are affected. However, when assessing
China as a context for the emergence of innovation, previous contributions tend
to assess current day developments from a limited contemporary perspective,
partly ignoring the insights that can be gained from taking a *historical* perspective on innovation as related to newly emerging phenomena such as the one
outlined above. Although the emergence of China has attracted much attention
among scholars and practitioners in recent years, from a historical perspective,
the emergence of new centers of innovation is not a novel phenomenon, as
earlier examples such as the United States in the 18th and 19th century suggest,
which produced a number of break-through innovations in areas such as transportation, energy and consumer goods. However, research taking a systematic
historical perspective of current innovation issues in China is currently lacking.

This dissertation seeks to fill this gap in the existing literature. First, it
provides a historical perspective on the economic, institutional-political and
sociocultural aspects of different innovation systems, focusing on the evidence
from innovation emerging between the First and Second Industrial Revolution,
and up to the period following the Second World War. Second, it applies the

insights gained from this historical analysis to the evidence from four case studies of Chinese and foreign multinational corporations (MNCs). In doing so, it shows not only *that* history matters for the study of innovation and management in general, but also *how* it does so (Jones & Khanna, 2006).

Overview: the following chapter 1.1 outlines the practical and theoretical motivation of this dissertation. First, the phenomenon underlying this research is described. Second, gaps in current research and existing theories are uncovered. Chapter 1.2 presents the research objective and the research question. The following chapter 1.3 defines the focus and definitions underlying this study. Chapter 1.4 provides an overview of the structure of the thesis.

1.1 Motivation

Much of the know-how of western multinational corporations in innovation rests on experiences from their home markets or other well-established R&D locations in Europe, the United States or Japan. However, apart from a lack of knowledge that often exists with regards to non-western geographic business contexts, scholars and practitioners in the field of innovation and business tend to ignore lessons learnt from the past, which can often render important insights for challenges and projections in the present and future. For example, while government support for R&D and innovation in contemporary China has been criticized, the postwar (i.e. Cold War) experience of substantial government spending on military related R&D projects in the United States, as well as other examples in other regions are often neglected.

The emergence of new global centers of innovation from a historical perspective may be another one. However, as these developments are relatively recent phenomena and continue to evolve, e.g. in the case of China, previous contributions have not sufficiently addressed this phenomenon from a more long-term, historical perspective.

While a plethora of media has often been quick at labeling new developments in China as "unprecedented", economic historians have long argued in favor of using historical approaches to analyze current-day events (e.g. Trompf, 1979). For example, two major theories of social change – cyclical and dialectic theory – suggest that a historical perspective allows for a more qualified, long-term view of current events. While the cyclical view proposes that changes in human society follow a pre-defined cycle – growth, development, and decay – dialectic theory predicts that social development follows a set pattern. In this context, Mark Twain is said to have noted, "history does not repeat itself, but it does rhyme" (Eayrs, 1971).

Due to the importance of innovation in China for theory and practice, as well as a lack of previous contributions using a historical perspective to increase understanding of related current-day phenomena, this dissertation applies a historical perspective to shed light on the economic, institutional and political as well as sociocultural aspects of innovation in China, applying a historical perspective on innovation to the empirical evidence based on four case studies of Chinese and foreign multinational corporations (MNC) in China.

The issues outlined above have strong theoretical and managerial implications, as companies from traditional innovation centers seek to benefit from conducting innovation in new markets, but need to adapt to new contexts. The following chapters provide further insights about these issues. Chapter 1.1.1 will outline in detail the practical relevance of the present research and provide current managerial challenges. Chapter 1.1.2 will exemplify several limitations of current theory.

1.1.1   Relevance of research subject

Gaining a better understanding of China as an emerging center of innovation based on historical context is highly relevant from a managerial as well as theoretical perspective.

The recent surge in innovation in China is mainly driven by rapid economic growth and the emergence of a large domestic consumer market, which provides enormous opportunities for the development and localization of products from foreign and domestic companies. In recent decades, Chinese consumers' income levels have increased dramatically, creating a rapidly growing middle class, which has resulted in a significant economic and social transformation process in China. According to a recent study, by 2022, more than 75 percent of China's urban consumers will earn 60,000 to 229,000 RMB ($9,000 to $34,000) a year. In terms of purchasing power parity, that range is between the average income of Brazil and Italy (Barton, Chen, & Jin, 2013). While just four percent of urban Chinese households were within this income bracket in 2000, in 2012, 68 percent of urban consumers were in that income range (Barton et al., 2013).

Within China, there is currently still a strong regional divide in terms of income levels between the highly developed coastal regions, and the more in-land Western regions. In 2002, 40 percent of China's still relatively small urban middle class lived in the four major ("tier-one") cities of Beijing, Shanghai, Guangzhou, and Shenzhen (Barton et al., 2013, p. 5). By 2022, the share of these large cities is predicted to decrease to about 16 percent, as the middle class is growing even more rapidly in smaller cities in northern and western China. The share of China's upper-middle-class households from these "tier-three" cities is expected to reach more than 30 percent by 2022, up from 15 percent in 2002 (Barton et al., 2013, p. 5).

This is to illustrate that while consumer demand has been booming primarily in the coastal regions of China, this trend is likely to continue towards the hitherto poorer regions in China's inland. Furthermore, previous contributions suggest that innovation-related activities often follow to those areas that previously experienced manufacturing and consumption-based growth. Such activities in China are likely to continue to expand also to these regions, as the evidence of increasing R&D centers in non-coastal regions suggests.

Furthermore, not only has innovation been shifting from developed to emerging markets. The nature of innovation activities is also changing. For example, in emerging markets like China, consumers are often resource-constrained, seeking "good-enough" (Christensen, 1997), more affordable products that fulfill their basic needs. Foreign MNCs in China have struggled to engage in the development of these lower-cost, functionality-centered innovations, due for instance to higher cost structures, as well as concerns over brand value.

In addition, emerging economies are no longer only receiving innovation, but they are increasingly also the origin of innovation. In 2014, China already had the second largest number of R&D centers worldwide, behind the United States and ahead of Germany, Japan and India (GLORAD R&D Database, 2014). Further, Chinese companies increasingly commercialize their innovations also in Western markets. One example is Haier, a manufacturer of home appliances, which has successfully expanded its business abroad. This kind of "reverse-innovation" challenges hitherto dominant innovation paradigms, as the existing organizational structure of western MNCs is often optimized for the development of advanced products and technologies targeted at high-end consumers (Widenmayer, 2012).

Thus, there is still a great amount of uncertainty among scholars and practitioners on a potentially emerging Chinese model of innovation and what this may entail. Innovation in China, strongly influenced by government policies and domestic cultural and social practices, and learning from western multinational companies, indeed provides an important direction for studies on innovation, e.g. considering the integration of efficiency-led business models in the West and effectiveness-led models in China. This dissertation seeks to shed light on these issues, by applying a historical perspective to the contemporary perspective on China's economic, political and sociocultural context of innovation.

1.1.2   Limitations in current theory and research

The topic addressed in this paper is relevant for scholars as well as practitioners, not only due to China's growing importance as a hub for R&D and innovation (Sun, Von Zedtwitz, & Fred Simon, 2007), but also due to a current lack of more comprehensive understanding of China's political economy, its institutional and social history and the related impact on innovation in Strategic Emerging Industries (SEI) in China, which a historical perspective can provide. In recent years, the institution-based view has been established as a third leading perspective in strategic management – besides the industry-based and resource-based views. As such, it has helped to overcome long-standing criticisms of the industry-based and resource-based views' lack of attention to contexts. It also contributes significant new insights as part of the broader intellectual movement centered on new institutionalism (Peng, Sun, Pinkham, & Chen, 2009).

However, currently there is only very limited understanding of how a historical perspective can illuminate our understanding of innovation in China. Indeed, historical study increases the robustness of a discipline, as it enables scholars within that disciplines, as well as society at large, to gain a better comprehension of its origins and its patterns of change. This kind of study relates a discipline to its own past and to other disciplines and therefore helps in establishing an identity for a discipline by providing some idea of *where it is* and *what it is* (Savitt, 1980, p. 52).

This research seeks to fill this gap, by providing insights on how the Chinese context for innovation differs from the one in Western economies from a historical perspective, thus informing the view on innovation in China for scholars as well as practitioners.

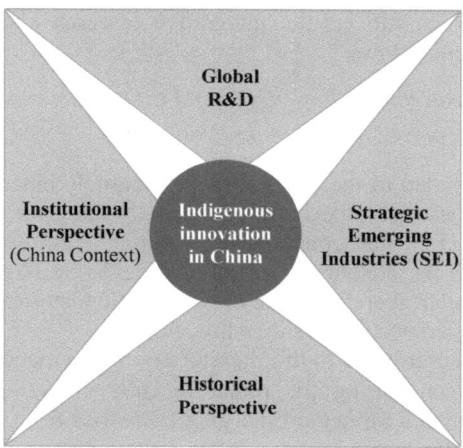

Figure 1: Overview of literature streams used in dissertation

1.2 Research objective and research question

The overall objective of this study is to increase understanding of the emerging phenomenon of innovation in China with a focus on innovation taking place within China's seven strategic emerging industries (SEIs). In particular, the aim is to better understand how a historical perspective on innovation can inform our current understanding of China as a hub for innovation.

The relevant literature on R&D management and global innovation has focused for the most part on providing analysis on currently emerging innovation patterns in China. However, as this dissertation argues, ignoring the broader historical dimension of innovation in China may result in a limited perspective and pre-empts the possibility of reframing our current perspective, for two main reasons. On the one hand, many developing countries like China have undergone dramatic political, economic and sociocultural changes in recent and also more long-term history. Therefore, an assessment of the current situation in China that does not take this into account is bound to be incomplete. On the other hand, when evaluating epochal developments such as the re-emergence of China as a leading economy in global innovation, a more long-term historical perspective is needed, considering how innovation has emerged in other historical settings, e.g. in Britain or the United States in the 18th and 19th century (more information about the historical approach will be provided in chapter three).

The main research question, which reflects the integrative research app-roach of this dissertation, is therefore as follows:

How can a historical perspective on innovation qualify and extend our eval-uation of current-day China as an emerging hub of innovation?

While some previous studies have alluded to the concept of a potentially emer-ging Chinese model of innovation, these have been rather patchy, calling for a more systematic analysis of those factors that constitute a potential "Chinese" model of innovation, as compared to the adaptation of currently existing (e.g. western) innovation models. In particular, there has been no systematic historical analysis on how an economic history perspective qualifies this view.

To summarize, this dissertation contributes to the literature on R&D mana-gement and global innovation, by considering the rather recently emerging phenomenon of firm innovation in China's Strategic Emerging Industries from a historical perspective, in order to extend current knowledge on innovation in China.

1.3 Focus and definitions

The following section outlines the focus of this dissertation and clearly frames the research subject. It further provides definitions of the key concepts covered in this dissertation, which will be further expanded upon in chapters three and four.

1.3.1 China as a research setting

In recent decades, several large emerging economies such as India, China, Brazil, and, to a lesser extent also Russia, have experienced high levels of econo-mic growth. Among those countries, the rapid growth of the Chinese economy has been most remarkable, with its Gross Domestic Product (GDP) growing at an average of 10 percent in real terms since the late 1980s. From a theoretical as well as practitioner's point of view, selecting China as a research setting for the purpose of this dissertation makes sense for five main reasons.

First, the large size and dynamism of the Chinese market provide enormous incentives for MNCs as well as domestic firms to develop and market innovation in China. In recent decades, China has undergone a dramatic period of economic transformation from a centrally planned to an increasingly market-based econo-my. Since the beginning of its opening reforms in 1978, initiated under the leadership of Deng Xiaoping, China's economy has experienced rapid economic growth. In parallel with China's increasing role in the global economy, China has become the largest exporting country in the world, and the second-largest importer (Eurostat, 2014). Among the European Union's largest single-country trading partners, China has become the second-largest export destination for European goods after the United States, and the largest source of imports entering the European Union. China's large and dynamic domestic market, featuring a population of more than 1.3 billion citizens and a growing middle class eager to purchase foreign goods, has also served as a stimulus for many European companies that sell goods and services to Chinese customers, and as a hub for manufacturing products that are sold in China and abroad.

Second, while many European companies have traditionally perceived China as a strategic location for low- and medium-value manufacturing, increasing wage levels in the manufacturing sector have reduced some of the cost advantages of moving production to China, with potential adverse effects on levels of foreign investment into China. Chinese government policies that are aimed at moving its domestic industry towards higher levels of innovation therefore also affect foreign companies operating in China. The speed and dimension of these processes occurring in China may therefore be of a historically unprecedented dimension, thus lending itself well to an analysis from a historical perspective as used in this dissertation.

Third, although MNCs have traditionally viewed China primarily as a location for low- and medium-value manufacturing, China is rapidly emerging as a large recipient of global R&D investment and a hub for innovation (von Zedtwitz, 2004). This is due to several reasons. Besides the attractiveness of the large domestic market, China features a very large potential talent pool for R&D activities. Every year, 2.5 million students graduate from Chinese universities, including 14,000 local PhD students (Gassmann, Beckenbauer, & Friesike, 2012). Due to lower labor costs in China, there are also cost advantages to developing innovation in China. Furthermore, as business success heavily depends on good relationships and networks (关系 or "guanxi" in Mandarin) in China (e.g. through cooperation with local governments, universities and research institutions), many MNCs have been involved in the Chinese market for a long time, increasing the availability of data and access to China-related managerial expertise.

Fourth, due to its large market and increased investment, in recent years China has rapidly developed into a major driver of global R&D. According to a recent OECD report, the rise of China, driven by its economic dynamism and its long-term commitment to Science, Technology and Innovation (STI) should continue. China's Medium and Long-term National Plan for S&T Development (2006-20) targets R&D spending of 2.5 percent of GDP by 2020. Assuming linear growth in Chinese and US R&D expenditure, China should outpace US R&D spending by about 2019 (however, China's recent economic slowdown may delay this scenario). The situation in the European Union will be more varied, and several countries will struggle to achieve a three percent target by 2020. The following figure shows how China might replace the US as the leading R&D spender in the coming years, in terms of gross domestic expenditures on R&D (abbreviated as GERD) (OECD, 2014, p. 58).

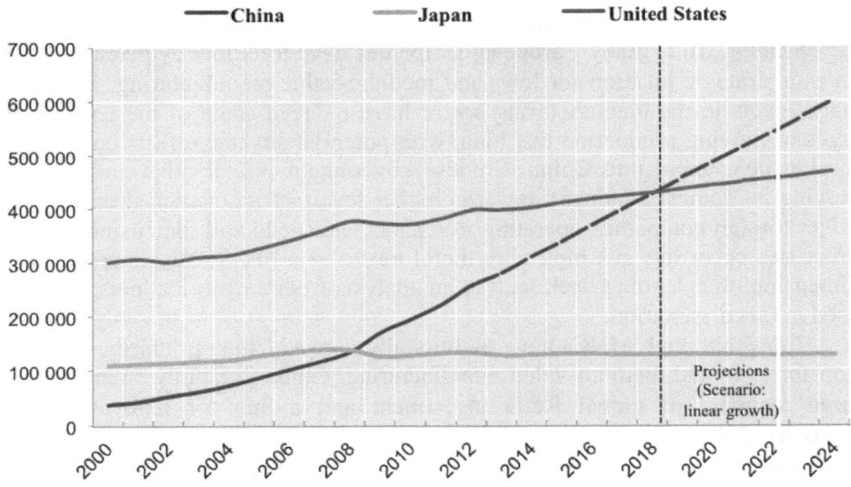

Figure 2: GERD, 2005 USDm (in PPP), 2024 projection (OECD, 2014, p. 58)

Fifth, the Chinese government has become increasingly involved in supporting domestic levels of innovation. It has made the development of innovative technologies – e.g. in the area of renewable energy products solar, wind or thermal energy – one of its priorities. Partly, this is also due to a need to find solutions to deal with environmental concerns that China's rapid development has brought about, e.g. in the form of severe air and water pollution as well as significant traffic congestion. From the perspective of policymakers, as well as the Chinese public, supporting means to achieve more sustainable development and to reduce the rampant and health-threatening environmental pollution in China is therefore a necessity rather than an option. A number of policies, regulations, laws and guidelines have been introduced at the central and local levels to support the domestic development of such technological innovation.

The factors outlined above make the Chinese innovation environment a complex, dynamic, and highly interesting context for a historical perspective contributing to innovation-related research.

1.3.2 Innovation in emerging and transition economies

Innovation plays an integral role for corporations in all industries, as companies seek to achieve and maintain competitive advantage (Porter, 1980). In recent years, the internationalization of R&D activities and the increasingly global development of technological innovation have led to a lively debate among scholars and practitioners. Evidence such as increasing numbers of patent filings in different industries taking place in emerging markets suggests that the geographic locus of product innovation is gradually shifting from the currently leading industrialized economies centers of innovation towards emerging and transition economies such as China (Bruche, 2009; von Zedtwitz, 2004). In this context, it is important to distinguish between "emerging" and "transition" economies.

The term "emerging market" generally describes an economy that is progressing toward becoming advanced, e.g. as measured by liquidity in local debt and equity markets and the existence of some form of market exchange and regulatory body. In monetary terms, an emerging market can be defined as an economy with low-to-middle per capita GDP that is transitioning towards developed-market status ("Financial Times Lexicon," 2014).

In contrast, the term "transition economy" refers to the political and economic reform process of a country that is moving from a centrally planned economy to a free market ("Worldbank Glossary," 2014). The latter term usually refers to countries in Central and Eastern Europe and the Former Soviet Union, as well as to China. Therefore, China qualifies as both an emerging and transition economy. This dissertation will consider in particular the institutional context factors that characterize China as a transition economy, based on its particular political and institutional context of innovation.

As businesses increasingly conduct innovation in non-traditional markets, they have found the particular economic, social and political characteristics of transition economies to influence their operations in subtle but pervasive ways. Prior studies have shown that the institutional context of firms indeed matters for innovation and that more developed institutions are positively correlated with technological innovation, while weaker institutional regimes – for example, in terms of weaker legal enforcement of intellectual property rights – in transition economies are generally found to have a detrimental effect (Boisot & Child, 1996; Child & Tse, 2001).

While some studies have focused on the opportunities of emerging markets as hubs for technological innovation (Breznitz & Murphree, 2011; Bruche, 2009; Zheng Zhou, 2006), others have also highlighted the challenges that emerging markets' institutional environments bring about, especially for foreign multinationals conducting innovation. Several aspects of transition economies such as China have been found to influence the development of technological innovation, such as the protection of intellectual property rights; the nature of China's market and competition landscape; government policies; and national culture (Keupp, Friesike, & von Zedtwitz, 2012; Yang, Liu, Gao, & Li, 2010; Zhu, Wittmann, & Peng, 2012). These, as well as further aspects, will be considered in more detail in chapter four.

1.3.3 Defining "innovation"

In order to identify innovation types, it is first essential to define a 'technological innovation'. Different disciplines including management, marketing, engineering and even economies provide unique ways of defining innovation (Garcia & Calantone, 2001, p. 112). This section defines important terms that are used in this dissertation. First, the concept of "innovation" used in this dissertation is defined, as well as different types of innovation as relevant for this study.

Distinction between "invention" vs. "innovation"

Before defining innovation, it is necessary to distinguish between "invention" and "innovation", as this difference is sometimes confused. "Invention" can be seen as the action of inventing something new, such as a device, service, or method. In contrast, "innovation" is the consecutive, combined process of invention followed by its (e.g. commercial) exploitation (Prud'homme, 2012, p. 20). The cycle considered "innovation" is only completed once the invention is applied, e.g. by being introduced to the market, and thus given a practical purpose (Prud'homme, 2012, p. 20). The following section defines the term of "innovation" more precisely.

The origin of "innovation"

Originally, the term "innovation" is derived from the Latin words novus (meaning "new") or innovare (meaning "to make new"), which demonstrates the novelty aspect of innovation as a concept.

From a historical point of view, the concept of innovation is often associated with the works of Joseph Schumpeter. In his much-cited book "*An inquiry into profits, capital, credit, interest and the business cycle*", Schumpeter (1934) identified innovation as a critical source of economic change, arguing that economic change revolves around innovation, the activities of entrepreneurs, and market power. In particular, he defined five types of innovation:

I. The introduction of a new product or a qualitative change in the existing product

II. Process innovation that is unknown in the industry

III. The opening of a new market

IV. The development of new sources of supply for raw materials or other inputs

V. Changes in industrial organization

Furthermore, in his book "Capitalism, socialism, and democracy", he established innovation as a foundation for economic development more broadly, by developing the concept of "creative destruction" (Schumpeter, 2013), which describes the disruptive process of transformation that such innovations entail. In his view, entrepreneurs introduce innovations to the market, which serve to destroy the value of existing companies and products in favor of new business concepts and thus contribute to economic growth and development. Furthermore, Schumpeter argues that technological innovation may create temporary monopolies allowing for above-normal profits, which give rise to competition, bringing profits back to their equilibrium level. Moreover, he views these temporary monopolies as a necessary incentive for firms and entrepreneurs to develop new products and processes.

Schumpeter proposed three distinct stages that still serve as a thought foundation for innovation scholars today. The first stage involves the technical discovery of new things or ways of doing things, which Schumpeter labels as 'invention'. The second stage is where innovation occurs, which he defines as the successful commercialization of a new good or service based on previous technical discovery or a new combination of new and old knowledge. The third stage – 'imitation' – describes the general adoption and diffusion of new products and processes to markets. These definitions had a lasting influence on subsequent scholars, as they clearly distinguish between several concepts surrounding innovation.

More recently, the 1991 OECD study on technological innovations (OECD, 1991, pp. 303–314) has provided a useful way to capture the essence of innovation from an overall perspective: "Innovation is an iterative process initiated by the perception of a new market and/or new service opportunity for a technology-based invention which leads to development, production, and marketing tasks striving for the commercial success of the invention."

This definition also addresses two important distinctions. First, the innovation process always comprises the technological development of an invention combined with the market introduction of that invention to end-users through adoption and diffusion (Garcia & Calantone, 2001, p. 112). Second, the innovation process is iterative in nature, which implies that the innovation process automatically includes the initial introduction of an invention to the market, as well as the reintroduction of an improved innovation. The iterative nature of this process leads to different levels of innovativeness and therefore requires a typology to account for different types of innovations. The OECD definition also refers to 'technology-based inventions'. Technological innovations are "those innovations that embody inventions from the industrial arts, engineering, applied sciences and/or pure sciences" including innovations from the "electronics, aerospace, pharmaceuticals, and information systems industries" (Garcia & Calantone, 2001, p. 112).

Therefore, according to Utterback and Abernathy (1978), products are developed over time, with initial emphasis on product performance, then emphasis on product variety and at the last stage an emphasis on product standardization and costs (Abernathy & Utterback, 1978).

1.3.4   Types of firm-based innovation

There is widespread agreement that innovation is become increasingly important to firms as well as national economies, as global markets are becoming ever more integrated and dynamic based on rapid changes in technological development. However, defining different kinds of innovation is more challenging, as there are a large number of models, classifications, definitions and frameworks outlining different types of innovation. The following section introduces a selection of concepts and models that have been dominant in literature.

One early innovation model by Kenneth Knight (1967, p. 482) differentiates between the following four types of innovation: product/service ; production process innovation; organizational structure; and people innovation. In the 1970s and 1980s, several contributions suggest administrative, technical, incremental, radical, product, and process types of innovation models from an organizational point of view (e.g. Bantel & Jackson, 1989; Daft, 1978; Damanpour & Evan, 1984; Damanpour, 1991; Evan, 1966). These models generally had a binary focus, meaning that they were pairing different aspects of innovations such as administrative versus technical innovation, product versus process, as well as radical versus incremental innovation.

Technical innovation here relates to new products, processes or services, while administrative innovation involves changes in the social structure of the organization (Evan, 1966) such as "policies of recruitment, allocation of resources, and the structuring of tasks, authority and reward" (Daft, 1978, p. 198).

Building on these earlier binary classifications of innovation – product-process; administrative-technical; and radical-incremental – in recent years, several concepts have been developed that integrate these categories, identifying several categories of innovation.

For instance, Cooper (1998) suggests a multi-dimensional model that integrates three of the binary innovation relationships outlined above, assigning each combination a dimension in his cube-like model (e.g. product-process as one dimension). Cooper argues that any kind of innovation can have some aspects of any of the six types of innovation.

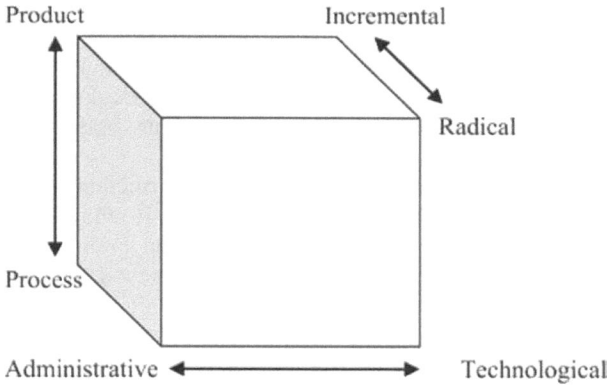

Source: Cooper (1998)

Figure 3: Cooper Model of Innovation (1998)

Similarly, other contributions have highlighted the relationship between innovation types (Boer & During, 2001; Rowley, Baregheh, & Sambrook, 2011).

Oke, Burke and Myers (2007) outline three innovation types: product innovation (radical and incremental); process innovation (comprising administrative, process innovation service and production); and service innovation. Therefore, they distinguish between product and service innovation. In their model, organizational innovation is a firm level innovation initiated by management.

Francis and Bessant (2005) define innovation based on the change that it brings about, proposing the following four categories of innovation (Bessant & Tidd, 2007):

Product innovation: changes in products and/or services offered

Process innovation: changes in how products and/or services are created/delivered

Position innovation: changes in context in which products/services are introduced

Paradigm innovation: fundamentally new ways of thinking about firm activities

In relation to the topic of this dissertation – innovation in China – the last two innovation types outlined by Bessant and Tidd (2007) are particularly relevant. 'Position innovation' occurs when firms explore new markets, customer bases and ways of serving them, e.g. in markets like China, or with simplified products targeting lower income customers. Similarly, 'paradigm innovation' can be highly relevant in the context of China, as it implies that firms may reframe their understanding of their products and services and thus may create markets that did not previously exist, e.g. in the case of low-cost airlines that are addressing new customer segments.

Recent contributions on innovation types have sought to further integrate the models and frameworks previously developed. For example, Rowley et al. (2011) have developed an innovation-type mapping tool based on Francis and Bessant's (2005) classification of innovation types, in which they develop a framework that distils and integrates the main types of innovation highlighted by previous contributions.

This dissertation assumes an open perspective on innovation, taking into account all types of innovations firms can potentially engage in, such as product, process, location and paradigm-based types of innovations. In order to operationalize the various innovation types outlined above at a firm-level perspective, the following section outlines different innovation types that can be differentiated based on three criteria:

A. *The degree of change*

B. *The origin/location of innovation (firm internal and external)*

C. *Innovation defined as a firm's strategy relative to other firms*

A. Innovation type based on degree of change

In general, previous contributions have considered innovations based on their degree of novelty, differentiating between "breakthrough" (also labeled as "radical") innovations for products at early stages of diffusion, and "incremental innovations" at the advanced stages of the product life cycle (Garcia & Calantone, 2001, p. 112). However, one may argue that "radical" as compared to "incremental" innovations may not form innovation types on their own, as they are rather attributes of any type of innovation than innovation types themselves (Rowley et al., 2011).

"Breakthrough" or "radical" innovations" refer to innovations that exploit existing forms of technologies, resulting in completely novel and cutting-edge innovations and often addressing new customer segments. One example is digital imaging technology used in consumer and professional cameras today, representing a radical departure from chemically coated film technology upon which George Eastman built the Eastman Kodak Corporation more than one century ago (Luecke, 2003).

In contrast, "incremental" innovations improve current innovations, but in a less dramatic way than radical innovation does. These innovations either improve something already existing or reconfigure an existing form or technology to serve a different purpose (Luecke, 2003). Therefore, incremental innovation is usually less risky and time-consuming than radical innovation.

It is important to note that both of the types of innovation outlined above are valuable. Indeed, within industries, incremental and radical innovations often go hand in hand. Over time, innovation is characterized by long periods of incremental innovation, interrupted by infrequent radical innovations (Luecke, 2003). For example, in the electronics industry, one can observe the introduction of radical innovation (e.g. large-screen televisions), followed by a series of incremental innovations (improving or reducing the price of a new product), and at some point again a radical innovation (e.g. the introduction of digital television).

A balance between different types of innovation is important, although radical innovation typically results in higher levels of competitiveness to the innovator. For this reason, as this dissertation highlights, governments in emerging economies such as China have used a variety of measures to increase levels of home-grown ("indigenous") incremental as well as radical innovation, in order to boost domestic companies' levels of technological sophistication and competitiveness (more details about such policies will be provided in chapter three). It is important to note that both radical and incremental innovation can occur at any stage in a firm's value chain, e.g. relating to innovation affecting product design or business models. The following section outlines innovation types originating from within the firm, as well as from the firm's external environment.

B. Innovation type based on the origin/location of the innovation

When thinking of innovation, the first association that usually comes to mind is product innovation. However, it is important to distinguish between outcome and process, and to locate the origin of innovation precisely in order to develop a more nuanced view of innovation, which includes a number of innovation types. These differ based on where the innovation occurs. They can take place at different stages along the value chain, as well as originating from within the firm ("inside-out" innovations), as well as from the firm's environment ("outside-in" innovations). Although the concept and sources of innovation are dynamic, with new types of innovations occurring over time, it is possible to group the most significant types of innovation based on the source of innovation within a product or process, providing an enterprise-focused perspective on innovation types for scholars as well as practitioners. According to the "Ten Types of Innovation" model developed by Keeley, Pikkel, Quinn and Walters (2013), from a firm-internal ("inside-out") perspective, innovation stems largely from process-related activities, as well as the quality of the product or service. The illustration below shows these ten types of innovations developed by Keeley et al. (2013).

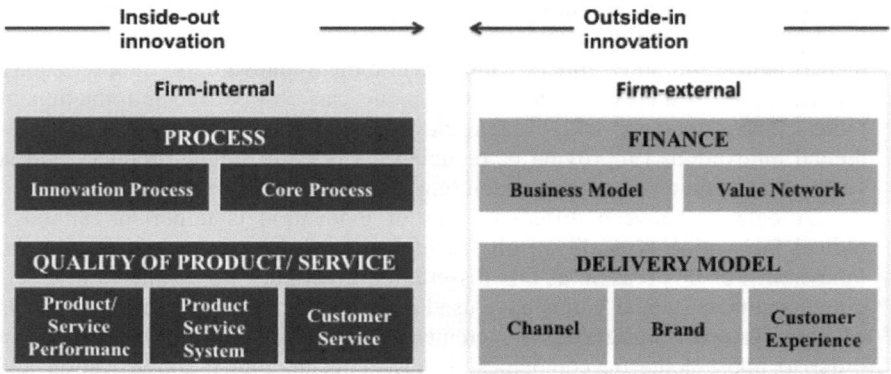

Figure 4: Innovation types based on Keeley, Pikkel, Quinn, & Walters (2013)

In the *inside-out perspective*, regarding *process*, such kinds of innovations can be attributed to firms using superior or new ways of doing business. They can refer to superior ways of managing the innovation process, as well as core processes of the firm, e.g. in production. Therefore, process-related innovations can occur when firms depart from "business as usual", which enables them to use unique capabilities, function efficiently, adapt quickly, and build market-leading margins. Often, process innovations form the core competency of a firm, including patented or proprietary approaches that result in sustained competitive advantage (Keeley et al., 2013).

Second, quite intuitively, the *quality* of the product or service can be another source of innovation. Innovations based on product performance relate to the value, features, and quality of a company's offerings. Product service system innovations are based on firms' ability to connect or bundle products and services to create a robust and scalable system, e.g. through interoperability, modularity, integration, or other ways of creating connections between otherwise distinct and disparate offerings. Lastly, *customer service* is another potential source of innovation, based on superior ways of supporting and amplifying the value of products and services in a way that increases the utility, performance and perceived value of the offering (Keeley et al., 2013).

From a firm external, *outside-in perspective*, innovation stems largely from activities related to *finance* and the *delivery model*. First, regarding finance, firms can attain high levels of innovation from choosing or developing an innovative *business model*, helping them convert their offerings and other sources of value into profits. This requires a deep understanding of customers and knowledge about new opportunities for revenue, as well as a willingness to challenge current industry assumptions. Firms that are first in developing new and profitable business models often enjoy sustained competitive advantage. Innovation in *value networks* is another type of innovation, stemming from the ability to cooperate with other firms to create value based on mutual benefit. One example of this is Apple's platform App Store selling third-party applications and software.

Second, regarding the *delivery model*, innovation can arise from using channels to connect the company offerings to customers and users in superior ways. Examples include innovative sales channels such as e-commerce (e.g. firms like Amazon) as well as innovative physical store concepts that attract more customers (e.g. IKEA). *Brand-based* innovation is another type, ensuring that customers and users recognize, remember, and prefer firm offerings to those of competitors or substitutes, thus creating a distinct identity and customer loyalty. Lastly, from the perspective of companies, *customer service* is another potential source of innovation that results in better relationships and loyalty with customers.

In addition to these specific types of firm-internal or external innovations, companies can also combine innovative solutions along patterns that reflect their innovation strategy. For example, firms seeking to develop innovation in an emerging market context often recombine or adapt existing process or product innovations, potentially resulting in innovative products targeting dynamic customer segments such as 'bottom of the pyramid' customers (Prahalad, Di Benedetto, & Nakata, 2012).

C. Innovation as defining a firm's strategy relative to other firms

Another perspective on firm-level innovation is to consider patterns of firm activities that shape *firms' strategic innovation objectives*. For example, in recent years, firms operating in emerging markets such as China have increasingly engaged in the development of "good-enough", no-frills products that initially respond to the demands of local customers, with the potential to be marketed globally (Gadiesh, Leung, & Vestring, 2007). Rather than being innovations per se, these approaches follow firm strategies and recombine particular firm innovations as outlined above, such as adapted business models, product design or new customer bases. One example is Mindray, a Shenzhen-based medical equipment manufacturer that has developed simplified electro cardiography (ECG) devices available at a fraction of the usual price. The following section introduces three kinds of innovation patterns that are especially relevant in the context of innovation in emerging markets like China, the focus of this dissertation.

1.3.4.1   Disruptive, frugal and reverse innovation

The term 'disruptive innovation' was shaped by Clayton Christensen in his article "Disruptive Technologies: Catching the Wave" (Bower & Christensen, 1995) and refers to innovations that typically emerge in simple applications at the bottom of the market and then move up market, ultimately replacing earlier technologies and driving out previous market leaders (Christensen, 1997). Examples of disruptive innovation include smartphones, which are increasingly replacing personal computers, discount department stores, as well as point-to-point ("no-frills") airline carriers.

This type of innovation occurs in a process in which innovations are initially inferior to mainstream technologies in terms of performance and only become disruptive once the established players deliver a "performance overshoot" (Yu & Hang, 2010), which over-serves customers, with the new disruptive product displacing the mainstream products.

According to Christensen, disruptive technologies underperform established products in mainstream markets. But they have other features that a few fringe (and generally new) customers do value. Products based on disruptive technologies are typically cheaper, simpler, and, frequently, more convenient to use. Therefore, disruptive innovations serve to make products more affordable and convenient and available to a much larger population, rather than being breakthrough technologies that make good products better.

Thus, rather than requiring a significant leap in technological refinement, disruptive innovations make changes to a product in unexpected ways, e.g. by targeting new consumer segments or by offer a product or service at a significantly lower price. Christensen defines disruptive innovation as being in contrast to sustaining innovation, which does not create new markets or value networks, but rather evolves existing ones with better value, allowing firms to compete against each other's sustaining innovations, with the latter innovations being defined as either discontinuous (i.e. transformational or revolutionary) or continuous (i.e. evolutionary).

Previous studies have found that the emergence of disruptive innovation is often based on the development of pragmatic solutions and adaptation as 'good-enough' products, rather than solely on technological superiority. For this reason, this type of products – also labeled "frugal innovation" (Zeschky, Widenmayer, & Gassmann, 2011), or "resource-constrained innovation" (Ray & Ray, 2010) - has been in great demand by the growing lower- and middle-class segments of consumers in transition economies such as China.

While a number of studies on disruptive innovation have critically assessed the theory of disruptive innovation (Danneels, 2004; Tellis, 2006) and have sought to further refine its theory and definition (Adner, 2002; Benner & Tushman, 2003; Markides, 2006), scholars are only starting to consider the impact of context and environment on the emergence of disruptive innovation (Yu & Hang, 2010).

However, although there seems to be great demand for products emerging from disruptive innovation in China, there is currently a lack of understanding of how the context of China influences the generation of disruptive innovation. Nevertheless, as mentioned above, the development of innovative products in emerging economies that initially serve a limited (e.g. lower-tier) part of the market and which may later on be marketed globally has become an increasingly relevant topic for both domestic as well as foreign companies operating in emerging markets. This concept of innovation can be subsumed under the name of "reverse innovation", introduced in the following section.

1.3.4.2   Reverse innovation

The term 'reverse innovation' was largely popularized by Vijay Govindarajan (Govindarajan & Ramamurti, 2011; Govindarajan & Trimble, 2009) and describes innovations that initially emerge in a developing economy and which are later introduced to an advanced economy. One example for this is the experience of General Electric, which introduced a simplified electrocardiograph portable device in the United States priced at only 20 percent of comparable products by competitors. This machine had originally been developed by General Electric's healthcare division to serve doctors in India and China.

The concept of reverse innovation challenges conventional models of innovation, as the latter are usually based on the idea that innovation flows move from developed to developing countries. For example, the well-established product life-cycle theory, developed by Raymond Vernon (1966), describes five stages of the product life cycle:

- Introduction
- Growth
- Maturity
- Saturation
- Decline

This theory, as well as other traditional views on innovation, assumes that new products and technologies are first developed in advanced economies, and only later introduced and commercialized in less developed economies, when they have become increasingly mature and redundant (von Zedtwitz, Corsi, Søberg, & Frega, 2015).

The notion of innovation originating in non-advanced economies was already established earlier, e.g. by Brown & Hegel (2005), who used the term "blowback innovation" to describe innovative solutions developed and adopted first in emerging markets. Hart and Christensen (2002) also applied the disruptive innovation framework to new products coming from developing countries.

However, there is still a significant lack of clarity in the defining this term, also differentiating it clearly from related types of innovations including disruptive innovation, innovation at the bottom-of-the-pyramid, indigenous, frugal or resource-constrained innovation, as introduced earlier (Widenmayer, 2012). In addition, earlier definitions of frugal innovation focus merely on the location of adoption and marketing of the product, excluding the location of the concept and development phases of innovation.

In a recent contribution, Max von Zedtwitz et al. adopt a linear model to conceptualize reverse innovation, which includes four sequential phases: *"concept ideation, product development, primary target market introduction,* and subsequently *secondary market introduction"* (von Zedtwitz et al., 2015). This typology builds on previous contributions on reverse innovation (e.g. Widenmayer, 2012), which have focused on market-introduction, integrating reversals in the flow of innovation in the ideation and product development phases. Analyzing geographical parameters for each of the four sequential innovation phases outlined above, 16 different types of innovation flows between advanced and emerging economies are presented, 10 of which can be defined as reverse innovation flows (von Zedtwitz et al., 2015). The resulting framework provides a more consistent terminology and an analytical model for studying global innovation and R&D patterns in general, and reverses innovation flows in particular.

In addition, it results in several implications for management. For instance, the study shows that firms that are able to manage their subordinate organizational units in a way that increases the potential for reverse innovation, are also more successful in their overall global innovation performance (von Zedtwitz et al., 2015). These insights are especially relevant for firms operating in emerging markets like China, as the ability to reverse innovate from China to other markets seems to be correlated to overall innovation performance. The following section outlines the unit of analysis that this dissertation is based on: western and Chinese companies operating in Mainland China.

1.3.5 Unit of analysis: European and Chinese corporations

The unit of analysis in this dissertation is Chinese and Western multinational corporations (MNCs). MNCs, also called multinational enterprises (MNEs), can be defined as corporations that have assets, such as facilities of production, in at least one country other than their home market, its headquarter location. Western MNCs refer to those corporations that have their headquarters in Europe or the United States. Chinese MNCs in this dissertation are defined as those that have their headquarters in Mainland China (excluding Hong Kong). This is regardless of ownership: for example, a western company owned by Chinese stakeholders would still be considered a western company, and vice versa.

Furthermore, ownership may be private, state or mixed (OECD, 2011). In China, relative to Western economies, a larger share of companies – both multinational and domestic – are partly or fully government owned, 'state-owned enterprises' (SOE). The term SOE refers to "business entities established by central and local governments, and whose supervisory officials are from the government" (Lee, 2009, p. 5).

With the foundation of the People's Republic of China in 1949, the government was the main actor in establishing and owning all businesses. Starting with the economic reform period in the 1980s under Deng Xiaoping, state-owned enterprises were increasingly restructured and in the following two decades, the government privatized many smaller and mid-sized SOEs, resulting in a number of initial public offerings (IPO). In 1978, SOE represented 77.63% of overall industrial production, with almost the entire remaining industrial production coming from collective-owned enterprises (Lee, 2009, p. 6). According to the World Bank, since 1999, the share of SOEs has declined from 37 percent to less than 5 percent in terms of numbers, and from 68 percent to 44 percent in terms of assets, most of which is due to SOE reform, in which smaller SOEs were either privatized or filed for bankruptcy to move them off government balance sheets (Worldbank, 2010).

Today, SOEs in China are often seen in what are regarded as strategic or sensitive industries such as energy and infrastructure often requiring large investments. Some of the largest SOEs in China today include SINOPEC (oil and gas), PetroChina (oil), Sinochem (chemicals), China National Offshore Oil (CNOOC), Sinofert (fertilizers), Bank of China, Zhejiang Expressway Company (infrastructure), and Dongfeng Motors (automotive).

As this dissertation seeks to include the historical perspective on contemporary issues surrounding innovation in China, the following section introduces the historical approach to innovation applied in this dissertation.

1.3.6 Innovation in China – integrating the historical perspective

As outlined above, in recent years, China has become one of the largest contributors to global R&D spending and this trend is expected to continue, with its share in global R&D spending continuing to increase. According to a recent study, R&D spending reached over US$ 280 billion in 2014, up 22% from 2012. According to this study, by 2019, China will overtake Europe measured by R&D spending and also overtake the United States by the year 2022 (Grueber & Studt, 2014).

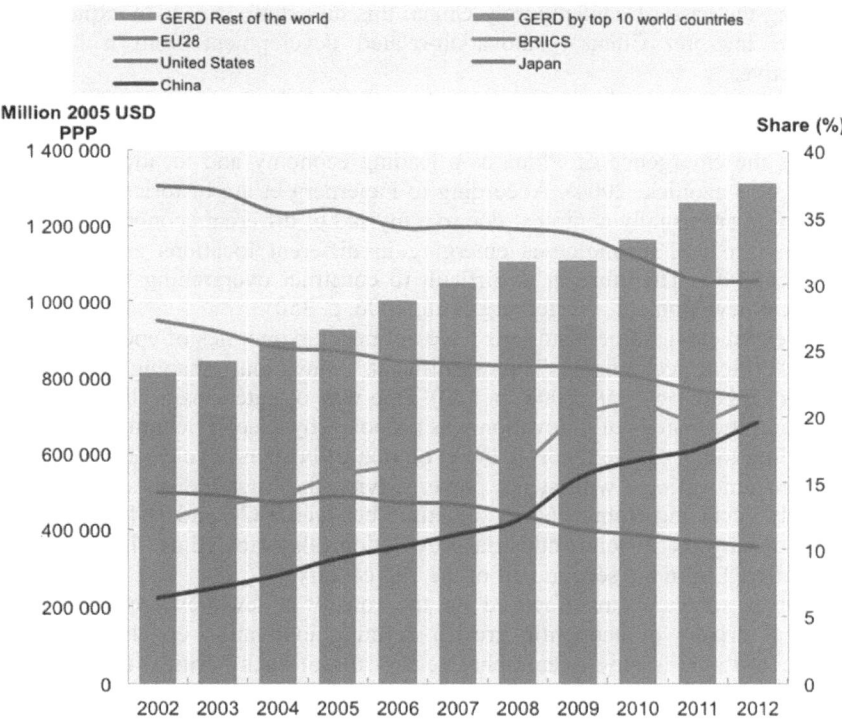

Figure 5: Share of top players in global R&D spending (OECD, 2014)

This development is due to increased foreign investment into R&D in China, as well as domestic investment, which has been encouraged in recent government initiatives such as the latest 12th Five-Year-Plan (China Central Government, 2011), which encourages China's further transition towards an economy driven by domestic consumption rather than exports, as well as higher levels of innovations, with domestic industries moving up the value chain based on higher levels of innovation. Due to the enormous size of its economy, the dimension of China's economic, political and social transformation process are unprecedented in history.

However, innovation-based economic transformation processes towards global leadership in general are certainly not unknown in history. Based on unique context conditions, the emergence of Britain and the US as global centers of innovation and economic leadership can serve as examples that are illustrative of this.

Studying the case of contemporary China, this dissertation seeks to explore how we can interpret China's innovation-related development from a historical perspective.

As previous scholars have pointed out, it is indeed necessary to include a historical approach when seeking to understand long-term patterns of innovation such as the emergence of China as a leading economy and location for global R&D (W. Lazonick, 2004). According to Fagerberg et al., historical patterns of innovation are usually complex, due to a myriad of different economic activities taking place and technologies emerging in different locations and industries (2006, p. 349). Therefore, it is difficult to construct overarching "schemata of historical development" (Fagerberg et al., 2006, p. 350).

Nevertheless, some historians have defined "taxonomies of epochs", identifying "critical technologies" as significant innovations shaping entire time periods (Fagerberg et al., 2006, p. 350). One way of categorizing different time periods in the history of innovation was proposed by Joseph Schumpeter (1939). According to his "wave theory", the evolution of business cycles are marked by different innovations, with steam power driving the First Industrial Revolution (roughly occurring from 1760 to sometime between 1820 and 1840), and electricity driving the Second Industrial Revolution (also labeled as 'Technological Revolution') from the second half of the 19th century.

Other scholars have focused on the role of a few key technologies as principal drivers of economic growth defining innovation epochs. However, these contributions may overemphasize the role of such "critical innovations", while neglecting other instances of innovation that also matter. Instead, one needs to acknowledge the complex and multi-sectoral character of innovation, and to take notice of the coexistence of various innovation modes, institutional processes, and organizational forms (Fagerberg et al., 2006, p. 350).

This dissertation uses an innovation system approach to illustrate changes in the innovation systems of different time periods, as well as the underlying evolution of economic activity, relevant institutions, as well as underlying flows of knowledge, an approach that has been established by previous contributions (e.g. Fagerberg et al., 2006). Using such an innovation system approach is appropriate, as innovation does not occur in isolation, but in a unique context of economic, social, political, organizational, institutional, and other factors that influence the development, diffusion, and use of innovations.

In order to be able to interpret and locate contemporary China's innovation situation from a historical perspective, the analysis will proceed as follows. This study will use a historical perspective to better understand how innovation systems have evolved in the past, and apply this perspective to the constantly evolving context of innovation in China. Following Fagerberg et al. (2006), this study will focus on the time period of the "First Industrial Revolution" between about 1760 and 1850. Second, what has been labeled as the "Second Industrial Revolution" will be discussed, covering the time period from the late nineteenth to the early twentieth century, a period that was marked by the rise of organization-based R&D. Third, the historical analysis will focus on context factors driving what may be labeled as the Third Industrial Revolution in the decades after 1945 dominated by the experience of post-war United States, with innovation originating in public research institutions, as well as private firms. The rationale for choosing these historical case studies is outlined in section 2.3 in the chapter outlining the research design of this study.

This analysis adds to the current knowledge about China as a rising economic and innovation center by using the historical perspective. For this, I will examine how factors that scholars of economic history have found to drive or impede innovation and economic development can apply to China.

The envisioned result of this analysis is a revised framework of context factors driving innovation in China that takes into account factors that have been decisive from a historical perspective, relating them to factors identified in the four case studies about Bayer Material Science, BYD-Daimler, Haier, and Siemens. The following illustration shows the initial framework used in the historical analysis section of this thesis.

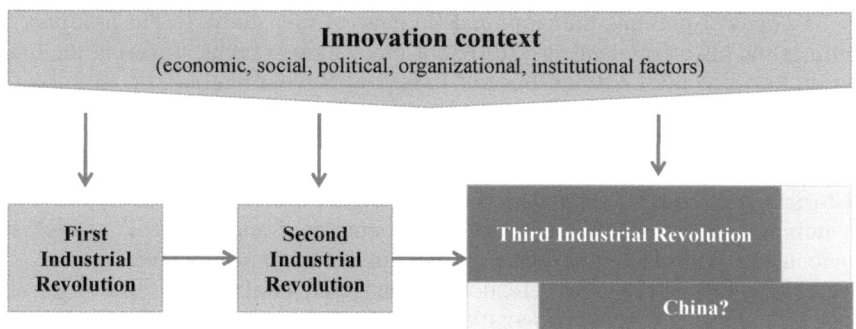

Figure 6: Initial research framework for historical analysis

1.4 Dissertation structure

Chapter 1 has outlined the theoretical and practical motivation of this dissertation, highlighting the need to integrate the historical perspective when evaluating contemporary China as a source of global innovation of rapidly increasing scale. In this chapter, the research gap is identified and the research question is presented. It further provides the focus of this study, as well as relevant definitions for the main issues addressed in this research.

In *Chapter 2*, the research design, concept, methodological approach, the use of data sampling as well as the underlying reasoning are presented. It further outlines the process of data collection and analysis.

Chapter 3 provides a theoretical understanding of the historical perspective on innovation, by outlining the most salient economic, political and institutional as well as sociocultural aspects that have been established in the literature. This is followed by a brief summary of previous contributions on innovation in China's Strategic Emerging Industries (SEI). Further, this chapter presents the most relevant literature global R&D management and the institutional perspective on China. Based on the theoretical insights, an initial reference framework is developed that guides the subsequent analysis on how our current perspective on innovation in China's SEI can be enriched by integrating the historical perspective.

Chapter 4 lays out the practice relevance of the historical perspective on innovation in China, by outlining the increasing relevance of innovation in innovation for practitioners, as well as the value of re-assessing relevant issues from a historical perspective.

Chapter 5 presents the empirical findings of this study. In the first part, it outlines the historical context of different innovation systems, covering the time period between the First and Second Industrial Revolution, and up to the time after the Second World War. In the second part, it presents the four in-depth case studies, Bayer Material Science, BYD/Daimler, Haier, and Siemens.

Chapter 6 synthesizes the insights gained, by applying the findings from the historical perspective on innovation to the insights from the case study analysis. In doing so, it considers the economic, institutional and political, as well as sociocultural aspects of innovation in China in historical perspective.

Chapter 7 outlines theoretical and managerial implications resulting from this study. First, it outlines propositions to extend the existing literature on innovation, which is enriched by the historical perspective. Second, it presents managerial implications resulting from a more nuanced understanding of innovation in China's Strategic Emerging Industries from a historical perspective.

Chapter 8 summarizes the theoretical and managerial implications. In addition, it points out to the limitations of this study and provides an outlook for future research.

The figure below provides an overview of the dissertation structure.

STRUCTURE OF DISSERTATION			
Introduction			
Motivation	Research Objectives Research Questions	Focus & definitions	Dissertation Structure
Research Design			
Research Concept	Research Method	Sample Selection	Data collection Data analysis
Historical perspective on innovation in China			
Historical Perspective	Innovation in China's SEI		Research Framework
Practical Relevance			
Innovation in emerging markets		The relevance of history for management	
Empirical findings			
The historical approach to innovation		In-Depth case studies: Innovation in China's SEI	
Synthesis: integrating the historical perspective			
Economic context	Institutional & Political context		Sociocultural context
Theoretical and managerial implications			
Theoretical implications		Managerial implications	
Conclusion			
Summary: theoretical implications	Summary: managerial implications		Limitations & further outlook

(Sections numbered 1–8 along the left margin)

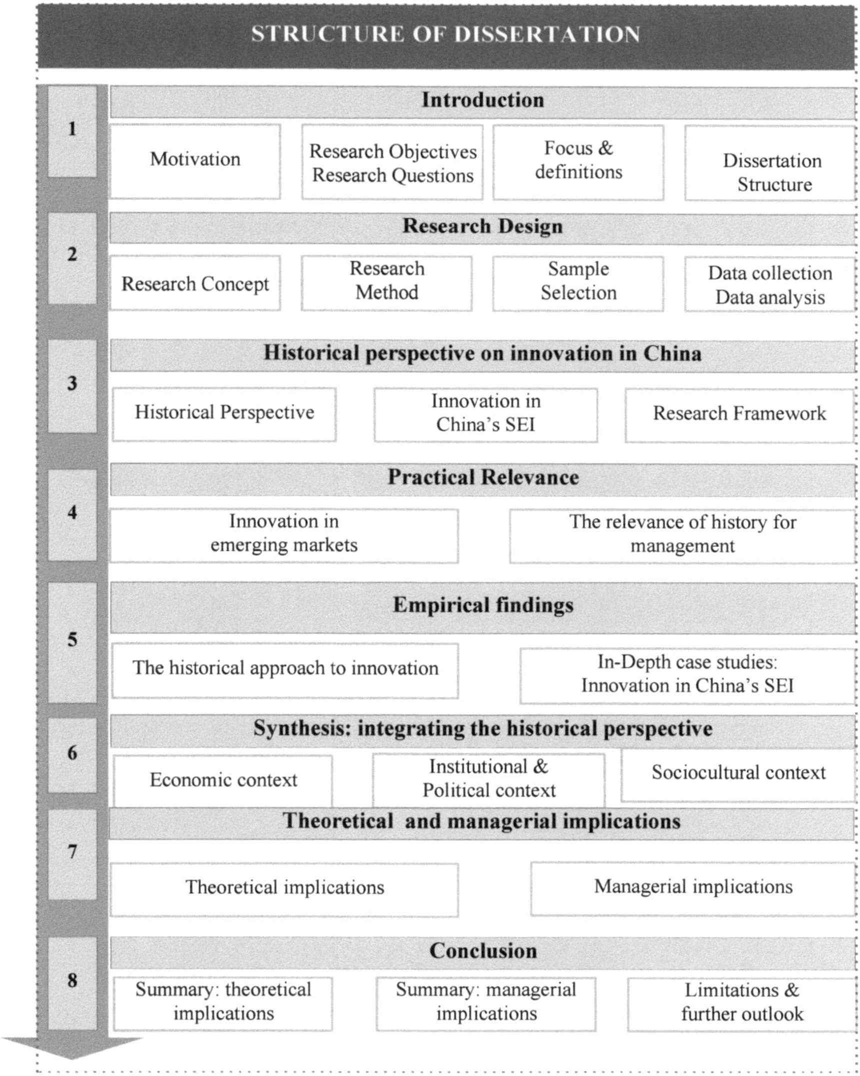

Figure 7: Overview of dissertation structure

2 Research Design

Overview: Chapter 2.1 presents the general goals and a generic research concept of this dissertation, as well as underlying assumptions. Chapter 2.2 provides the rationale for the case study method used for explorative research. Chapter 2.3 gives in-depth reasoning to the selected sample. Chapter 2.4 shows how empirical data was acquired and subsequently analyzed.

Research designs define the type of research study, its research questions, hypotheses, variables, methodologies used, and data collection methods. The research methodology should be selected with the purpose of maximizing research validity (Black, 1999; Yin, 1989). The choice of research design and methodology depends on the research question. The research question of this study is explanatory in nature, as it seeks to identify background information on *how* specific phenomena take place, and how this affects decision-making. This dissertation applies an explorative approach. To recall the research question:

> *How can a historical perspective on innovation qualify and extend our evaluation of current-day China as an emerging hub of innovation?*

In particular, this dissertation seeks to provide a better understanding of the emergence of innovation from a historical perspective, and applies the resulting insights to the context of innovation in Strategic Emerging Industries in China today. Therefore, the explorative approach is pertinent to the research focus on the discovery of relevant relationships and results in theoretical and practical implications. Apart from the historical analysis, firm-level evidence is collected and analyzed based on the relevant scientific literature and evaluated based on the results of the historical analysis.

2.1 Research concept

The present research is targeted at establishing a better understanding of emerging market innovation – with an empirical focus on China – from a historical perspective. Despite its high importance for management and scientific literature, empirical research to date has not yet critically assessed the burgeoning phenomenon of R&D and innovation in China from a historical perspective. Thus, while some previous contributions use historical anecdotes and evidence

peripherally, a thorough historical perspective is currently missing. This dissertation seeks to fill this gap in the literature and is rooted in two literature streams.

First, it builds on the literature on the history of innovation, illuminating the
evolving nature of innovation from a historical perspective and focusing in
particular on economic, institutional and political, as well as sociocultural
aspects of an evolving innovation context. This serves the purpose of expanding
existing perspectives on innovation in emerging markets, and particularly in
China, by including historical elements that can help inform our present views,
an aspect that has been almost completely neglected in previous contributions.

Second, it applies this historical understanding of innovation to the current-
day phenomenon of innovation in China, focusing on comparable economic,
institutional and political, as well as sociocultural aspects the Chinese innovation
context, building on relevant literature on global R&D and innovation management, as well as institutional theory.

This dissertation uses historical methods to understand innovation in the
past, as well as field research to understand innovation in present-day China. It
thus aims at contributing to existing literature and theory and assumes that research is an iterative learning process that can create knowledge based on theory
and practice, rather than validating hypotheses solely derived from theory (Berg
& Lune, 2004; Bonoma, 1985; Eisenhardt, 1989). Based on data analysis, observable elements and their interrelations are revealed, reflecting empirical data on
theory with new perspectives of reality by using differentiation, abstraction, and
changes in perspective, thus enhancing alternations and theory building (Skorna
& Widenmayer, 2010; Widenmayer, 2012, p. 32). Throughout the research process, empirical data is connecting and disconnecting with existing literature,
resulting in theory expansion (Mintzberg, 2005). Finally, as a result of the research process, propositions are developed that extend existing theory.

2.2 Research method

As outlined above, this dissertation follows an explorative research design, based
on the type of research question, as well as the multifaceted and novel character
of the phenomenon and relationships studied. This dissertation seeks to develop
theory from the application of historical research to selected case studies,
conducting an in-depth analysis of innovation history, and applying the resulting
insights to case studies, which follow a multiple-case design (Eisenhardt, 1989;
Yin, 1989). According to Yin (1989), a case study "is an empirical inquiry that
investigates a contemporary phenomenon within its real-life context, over which
the investigator has little control, especially when the boundaries between pheno-

menon and context are not clearly evident" (Jin, 2005, p. 49). Comparative case studies build theory in a post-positivist manner.

The use of historical research designs

Historical research in innovation can be seen as a narration of events through time in which their sequence is described. Historical research requires the "analysis and explanation of the causes and consequences of events with particular concern for change" (Savitt, 1980, p. 53) and it should explore and analyze the subtle relationships between historical innovation events and the underlying economic, political-institutional, as well as sociocultural context.

Historical research designs are used to gather, verify, and synthesize evidence from the past to establish facts that defend or refute a hypothesis. For this purpose, scholars can use historical evidence based on primary, as well as secondary sources. It is important to note that in historical research, it can be difficult to draw a precise line between primary and secondary sources of evidence (Savitt, 1980, p. 55). For instance, primary sources that are commonly used, e.g. historical public records, were collected by individuals at the time, rather than the history scholar. Primary sources typically used include government or other public records, corporate records, diaries, pictures, visual or audio recordings, and maps. Secondary materials include, but are not limited to, textbooks, journal articles, newspapers, biographies and other media such as films or tape recordings. Historical materials can provide important contextual background to increase understanding of a research problem. In order to achieve relevant results, the sources used must be both authentic and valid (Savitt, 1980; University of Southern California, 2014).

Historical perspectives can be defined as being *descriptive* and *comparative* (Savitt, 1980, p. 53).

They are *descriptive* because they begin as narratives of events in a specific time period, for example the period of the Industrial Revolution. Particular events are identified and described based on their specific characteristics. The subsequent analysis then provides explanations, relationships, and consequences of the events. In some cases, the insights gained from historical analysis can serve as a basis of prediction, when extrapolation to future cases is desirable and realistic (Savitt, 1980, p. 53).

Historical perspectives are also *comparative,* e.g. when events in a single place are compared over a specific time period, or when events at different places are compared in the context of chronological time – for example, at the same time, or over a specific period of time. While the former approach is part of traditional, chronological historical research, the latter example is also an example of historical study. Furthermore, it is important to note that the use of comparison is part of the historical perspective, even though it is only a method and not historical research itself (Savitt, 1980, p. 53).

Following the historical analysis, this study uses a *case study approach*, to increase understanding about innovation in Strategic Emerging Industries in China, based on the preceding historical analysis. Cases studies are in-depth and holistic inquiries of single or few incidents or cases of a phenomenon ("how" and "why" rather than "how much/often"). They are studies in their real-life context and may combine different methods (e.g. qualitative and quantitative) as appropriate. As a positivist and interpretive study type, they are generic research strategies aimed at producing memorable examples of important management issues and concepts.

Justifying the use of the case study approach

According to the relevant literature, the use of case studies as a qualitative research approach can be justified for this study based on the following main considerations:

First, case studies are used in particular when identifying contextual information on "how" specific phenomena occurred and what explains the resulting decision-making. Therefore, previous contributions suggest that the case study method is appropriate when 'how' or 'why' questions are being asked about a contemporary set of events over which the investigator has little or no control (Yin, 1989). This is true for the main research question addressed in this dissertation.

Second, choosing a case study approach is also sensible as it allows for the study of multifaceted phenomena, such as the emergence of innovation in its historical text, through a collection of data in natural settings, rather than relying on "derived data" (Bromley, 1986, p. 23). In addition, the using case studies allows for the integration of a variety of sources of evidence including documents, interviews and observations and thus strongly increases the comprehensive treatment and understanding of the topic (Yin, 1989).

Third, purely quantitative research approaches such as surveys typically involve a larger sample size, potentially resulting in higher levels of generalizability of results. However, the complexity and also sensitivity of some topics addressed and materials analyzed – in particular company level assessments of Chinese innovation-related government policies - excluded quantitative approaches such as online-based surveys. In addition, previous studies have shown that in Mainland China, managers typically prefer face-to-face interviews to questionnaires. Response rates for the latter thus tend to be very low in international comparison. Therefore, in China, it is often necessary to establish personal relationships ('guanxi') with respondents in order to receive qualified responses (Harzing, 2006).

Fourth, using a case study approach allows for interviews to be less structured than surveys in order to gain more comprehensive understanding of the topic, enabling the author to integrate discussions of issues arising spontaneously in the conversation, an important aspect in developing an understanding of new perspectives on theoretical phenomena as pointed out by Eisenhardt (1989) and Corbin & Strauss (1994).

Fifth, the multiple-case approach allows for comparison across sites, which reveals idiosyncratic characteristics of each site, increasing the robustness of findings (Miles, 1979) and in using this approach, this dissertation seeks to identify those characteristics and theoretical mechanisms that determine how firms conduct innovation in the context of China, based on a historical perspective.

Nevertheless, one of the main drawbacks of the case study method is the narrow and idiosyncratic representation of research results, which can make it difficult to provide an adequate generalization of insights (Jin, 2005, p. 49). To improve the validity and generalization of the research results of cases studies, a number of previous contributions have provided detailed guidance on how to perform case study research in order to minimize such drawbacks, and at the same time build on the unique strengths of the case study method in enabling the analysis of complex and multifaceted phenomena (Eisenhardt, 1989; Yin, 1989). The following sections describe how these recommendations were implemented in the present study.

Based on limited previous academic research, in-depth data on four case studies was collected. The cases were chosen deliberately on the basis of theoretical considerations following Eisenhardt (1989), as choosing the case studies randomly is neither necessary nor preferable and extreme examples are most appropriate when seeking to extend theory. Instead, it makes sense to deliberately choose cases, so that the "process of interest is transparently observable" (Eisenhardt, 1989).

Following Yin (2014), several measures were taken in order to increase the most important criteria for qualitative empirical research, validity (construct, external and internal) and reliability of results (Yin, 2014). Multiple sources of evidence were used including semi-structured material based on interviews, firm-internal documents, desk research and press clippings. Data triangulation was used to increase internal validity. Processed interview data was subsequently confirmed in follow-up interviews.

External validity confirms that the findings can be generalized within the frame of the conducted research (Yin, 2014). The data was processed to allow for a clear chain of evidence between questions asked, data collected and conclusions drawn. Concepts and theories emerging from data were compared with the literature for generalization and theory building from cases in order to increase internal validity of causal relationships (Yin, 2014). Lastly, reliability ensures that another researcher would be in the position to conduct the same research successfully, using the same procedure at a later date (Eisenhardt, 1989; Yin, 2014). It is therefore important to describe data collection and analysis in detail and in a transparent way. The following sections illustrate in detail how the quality criteria are addressed and how rigorousness of research is assured.

2.3 Sample selection

This section first outlines the rationale guiding the historical analysis, as well as the sample selection for the case study research.

Historical analysis: sample selection

As outlined in the literature review on historical drivers of innovation, innovation systems evolve over time and differ based on different industries and geographies. They are also marked by different speeds and levels of dispersion in different time periods and locations. Therefore, it is difficult to construct overarching schemata of innovation from a historical perspective that provide complete consideration to nuances and evolutions in the innovative environment over time. It is thus necessary to define a specific focus – e.g. based on time periods, technologies considered, or geography – in order to operationalize any potential historical perspective on innovation.

Indeed, a number of historians and innovation scholars have defined "taxonomies of epochs" (Bruland & Mowery, 2004, p. 1) to identify commonalities and patterns in the history of innovation, structuring their approach long different dimensions.

One way of doing this has been to center taxonomies around "critical technologies" that have defined whole periods. One example briefly mentioned above is the wave theory proposed by Schumpeter. In his two-volume, 1,095-page seminal work on "Business Cycles", Schumpeter (1939) develops this theory, building on earlier prominent business-cycle theorists including Clement Juglar, Joseph Kitchin, and Nikolai Kondratieff. In doing so, he focuses on five industries that led the process of economic development - cotton textiles, railroads, steel, automobiles, and electric power – and emphasizes three institutional innovations crucial to the rise of capitalism: the factory, the corporation, and the modern financial system. Other contributions have deprioritized the wave theory perspective, focusing instead on a few key technologies driving economic growth (Bruland & Mowery, 2004, p. 1).

However, it is easy to overstate the importance of individual, "critical" technologies while downplaying the role of other types of innovation that that also matter. This study seeks to move beyond the description of individual key technologies that have been seen as defining whole innovation epochs. It illustrates the evolution of innovation from a multi-sector perspective that takes into account different innovation modes, institutional processes, and organizational forms. In particular, the present research is based on an innovation system approach (the latter being outlined in the literature review section of this thesis) to describe an innovation context evolving in subsequent time periods. This approach shows the evolving nature of innovative activity, the underlying institutional framework, as well as underlying flows of knowledge in emerging industrial economies such as Britain and the United States in their respective time periods (Bruland & Mowery, 2004).

The historical analysis includes three distinct time periods: the First Industrial Revolution between about 1760 and 1850; the Second Industrial Revolution, which occurred from the second half of the nineteenth to the first half of the twentieth century; and lastly what some scholars consider to be the "Third Industrial Revolution" following World War Two, a time period that has been marked by the dominance of the United States, and increasingly also by the globalization of R&D and innovation (Fagerberg et al., 2006, p. 350). This encompasses the emergence of innovations based on the use of coal in Great Britain of the 18th century, the improvement of property rights and a greater awareness of the value of innovation as an end to itself over time and thus - in the words of philosopher A.N. Whitehead – an emerging notion of the "art of innovation" (Whitehead, 1925). In particular the third time period discussed, the time after the World War Two, illustrates the importance of private and public institutions in relation to innovation to this day.

The rationale for choosing these periods is provided in the following section.

Most importantly, it is important to note that invention and innovation did occur much earlier than the Industrial Revolution in the 19th century. However, as previous contributions have shown, a number of geographic, economic, institutional and sociocultural factors allowed for the Industrial Revolution to occur first in Britain, as a first instance of substantial economic growth on a broad basis resulting in increased productivity, embedded in an increasingly formal and supportive institutional environment (e.g. Crafts, 1977; Landes, 2003; Mokyr, 2010). Only the particular context of the Industrial Revolution provided an environment in which sustained innovation-based development could emerge (Bruland & Mowery, 2004, p. 2).

Thus, the selection of the three particular time periods and locations does not seek to understate the importance of other inventions and innovations in other time periods and geographies. Indeed, the historical and cultural context of Britain, Western Europe and the United States is remarkably different from the one encountered in China. While this may suggest that a comparison of China with the historical experience of other (e.g. Asian) and more similar economies would be most appropriate, the selection of the three time periods (and regions) outlined above is more suitable, as it allows for an evolutionary analysis of innovation systems in different time periods and regions that were interrelated and spread to other parts of the world including Asian economies.

Furthermore, the time period of the Industrial Revolution is well documented and numerous previous contributions have illuminated it from different perspective. This is helpful for this study, as it allows for a specific analysis of innovation from a historical perspective that can then be transferred to the new phenomenon under consideration, the context of innovation in China's Strategic Emerging Industries.

Indeed, it is difficult to speak of such kind of a coherent innovation system that includes different stakeholders such as individual inventors and innovators, government institutions and policies, and organizations such as companies before the time of the Industrial Revolution. Furthermore, the choice of three broadly related subjects of analysis – in terms of geography, as well as subsequent time periods - allows for a comparative analysis of innovation taking place in an evolving context of influencing environmental factors. The Industrial Revolution occurred first in Britain and then spread to continental Europe and the United States. Therefore, the subsequent two sections are naturally focused on these regions. As innovation subsequently - and in particular after the Second World War – also emerged in other regions such as Asia, the relevant sections also include important insights from innovation occurring in these regions.

Theoretical sampling: company case study selection

According to Eisenhardt, the case selection is a crucial element of building theory from cases (1989, p. 536). Similar to hypothesis-testing research, the concept of a population is important, as it defines the set of entities from which the research sample is to be drawn. Furthermore, the selection of an appropriate population controls extraneous variation and helps to define the limits for generalizing the findings. According to Eisenhardt (Eisenhardt, 1989), it makes sense to choose cases "in which the process of interest is transparently observable" (p.537). In theoretical sampling, one should choose cases that are likely to replicate or extend the emergent theory. The approach to sampling the firms analyzed in this paper followed Eisenhardt in that "the goal of theoretical sampling is to choose cases which are likely to replicate or extend the emergent theory" (Eisenhardt, 1989).

In contrast, traditional, within-experiment hypothesis-testing studies rely on statistical sampling, in which researchers randomly select the sample from the population (Eisenhardt, 1989). This is in line with the overall goal of this dissertation, which seeks to advance or develop existing or new theory instead of theory testing on a broad scale (Eisenhardt, 1989; Yin, 2014). As Pettigrew (1990) notes, given the limited number of cases which can usually be studied, it makes sense to choose cases such as extreme situations and polar types in which the process of interest is "transparently observable".

The case studies were selected through the following process.

A preliminary list of companies operating in Strategic Emerging Industries (SEI) in China was compiled, based on previous interviews conducted with 27 companies in China in the context of a research project in cooperation with *IP Key*, a three-year-project (July 2013 - June 2016) based in Beijing, China, implemented by the Office for Harmonization in the Internal Market in partnership with the European Patent Office (IP Key Project, 2015). This list of companies was then short-listed to those four companies with the largest potential for learning with respect to the phenomenon studied in this dissertation. To facilitate this process, additional interviews were conducted with representatives of two international intellectual property law firms (one headquartered in Beijing, China, the other one headquartered in Sydney, Australia), the German and US Chambers of Commerce in Shanghai, as well as with a senior editor of China Daily News in Beijing, representing China's leading English speaking newspaper.

The companies selected for the in-depth case study sample are Siemens AG, Bayer Material Science, BYD-Daimler, and Haier Group. While Siemens and Bayer Material Science are headquartered in Germany, BYD (based in Shenzhen) and Haier Group (based in Qingdao) are based in China. These four case studies were chosen based on the following criteria.

First, the companies are operating in at least one Strategic Emerging Industries (SEI) in China, as outlined earlier.

Second, the selected companies have an R&D unit in China and have built up significant R&D activities in Mainland China. Further, as the focus of the attention was mainly on China's innovation context rather than individual firm strategies, the headquarter location of firms was not a selection criteria. Instead, following Yin (1989), cases were selected as to be interesting and relevant to the study objective and to increase external validity to represent a large range of situations to which the study can be generalized.

Third, the companies have a global footprint in terms of sales, production and R&D, putting them in a position to consider and compare different R&D locations globally. They have established structures and a longstanding history in global management enabling global development.

Fourth, the companies are of sufficient size and leading players in their respective industries in China, potentially making them susceptible for the influence of Chinese industrial and innovation policies.

Fifth, the company headquarter location may be either in China or another country.

Sixth, using two cases from the same or related industry (i.e. Siemens and Haier) or with comparable histories (i.e. longstanding industrial corporations such as Siemens and Bayer), in addition to within-case analysis, it is possible to identify cross-case patterns and similarities and differences, which increases external validity and helps in breaking "simplistic frames" (Eisenhardt, 1989).

Seventh, all companies selected use technological as well as business model innovation to be industry leaders, building on their respective competitive advantages.

Although there are no specific scientific recommendations regarding the number of case studies in qualitative research, in the relevant literature, between three and ten are regarded as sufficient for theory development (Eisenhardt & Graebner, 2007; Widenmayer, 2012, p. 40). For this dissertation, four case studies were selected that offer interesting insights into the emergence of innovation in the context of China. The following table provides an overview of these companies. Detailed information about the case units is presented in Chapter 5.

Table 1: Selected case studies of dissertation (source: latest annual report)

Company	Industry sector	HQ location	Revenues (2013)	Employees	Distribution of revenues
Bayer Material Science	Polymer materials	Leverkusen (Germany)	EUR 11.2bn	14,300	Europe: 39% Asia/Pacific: 27% North America: 22% Latin Am/Mid-East: 12%
BYD-Daimler	Automotive	Shenzhen (China) / Stuttgart (Germany)	BYD: CNY 49.8bn (EUR 6.1bn) Daimler: EUR 118bn	BYD: 159,000 Daimler: 274,600	*BYD* China: 86% / India: 2% Europe: 2% / USA: 4% Other: 6% *Daimler* Western Europe: 35% NAFTA: 28% / China: 9% Other Asia: 12% / Other: 16%
Haier	Consumer Electronics Home Appliances	Qingdao (China)	CNY 180.3 (EUR 21.9bn)	55,800	n.a.
Siemens AG	Conglomerate (Industry, Health-care, Infrastructure)	Munich (Germany)	EUR 75.9	362,000	Europe: 53% Americas: 28% Asia, Australia: 20%

2.4 Case study data collection and data analysis

2.4.1 Data collection

The empirical research and data collection on the four case studies was conducted between September 2013 and December 2014. In terms of methods or research "tools", data were collected using semi-structured, in-depth personal interviews, archival documents, and on-site observations. According to Yin

(1989) and Eisenhardt (Eisenhardt, 1989), the triangulation of data gathered with different methods can increase construct validity, by reducing the limitation of using a single method, thus strengthening theory development. The following table provides information about the data sources for each of the four case studies.

Table 2: Data sources of the four case studies

Company	# Interviews	Interviews / Location	Archival documents
Bayer Material Science	5	5 in Shanghai, China (personal)	Annual reports; websites; business plans; patents; company presentations; organization charts; press releases; new clippings
BYD-Daimler	4	3 in Beijing, China (personal) 1 via telephone	Annual reports; websites; business plans; patents; company presentations; organization charts; press releases; new clippings
Haier	3	1 in Shanghai, China (personal) 2 via telephone	Annual reports; websites; patents; company presentations; organization charts; press releases; new clippings
Siemens AG	5	5 in Beijing, China (personal)	Annual reports; websites; business plans; patents; company presentations; organization charts; press releases; new clippings

Interviewees were either native German, Chinese or English speakers; for comparability reasons, all interviews were conducted in English. Respondents were chosen based on their experience related to R&D as well as the innovation context of China.

The author personally conducted individual, open-ended, and semi-structured (also referred to as 'focused') individual interviews. Semi-structured interviews can lead to more flexibility and the possibility to make adaptations; they can also create a more free and open atmosphere for the interviewer as well as the respondent (Yin, 2014). The interviews asked for both past and present data to create greater depth of understanding of how events had evolved over time.

The interviews were organized following the same semi-structured interview guide consisting of several open-ended questions that allowed the informant to relate his or her experience (the interview guide can be found in the appendix). In most cases, respondents were senior R&D managers. Personal interviews lasted between 60 and 130 minutes; phone interviews lasted between 40 and 90 minutes. At the end of each interview, interviewees were asked to name further colleagues who may be able to provide further information on innovation-related activities in China ('snowball sampling'). Due to the sensitivity of issues addressed in the interviews, the interview participants requested that their personal names be removed in the final study for confidentiality reasons.

The author tried to control for potential respondent bias by not discussing any elements of emerging theory with respondents and by maintaining a neutral presence during company visits and interviews. Further, to reduce bias from recall and rationalization, the author triangulated the collected interview data with both firm-internal data sources and external analyses from third parties, as provided in the table above. Using these procedures and additional materials was done in order to reduce potential respondent bias. Additional materials used for data triangulation included annual reports; websites; business plans; patents; company presentations; organization charts; press releases; and new clippings. Further interviews were conducted and materials gathered until there was no more marginal improvement of understanding and theoretical saturation was thus reached (Eisenhardt, 1989).

2.4.2 Data analysis

All semi-structured interviews were recorded with detailed interview notes. Due to the sensitive nature of information shared in the context of R&D facilities in China, it was not possible to tape record the interviews for subsequent transcription. The data received from the interviews were compared with the information received from non-personal sources. If discrepancies emerged, these were clarified with representatives of the specific company. Interview data were first entered into word processing software and subsequently entered into a case database, together with all other company-related data. Afterwards, this material was used to compile individual case histories of 10-15 pages in length including data from the interviews, as well as third-party information. While analyzing interview notes and other documents, the author iteratively compared the case materials with theoretical contributions to compare and relate emerging theory with the data, also by developing tentative propositions to describe emerging themes (Eisenhardt, 1989).

3   Historical perspective on innovation in China

Overview: This chapter outlines the theoretical concepts used in this dissertation, based on a review of the relevant literature, and outlines the theoretical relevance of the study.

In the first part, the historical perspective on innovation is presented. This includes an overview of factors that have been found to be conducive to from a historical perspective. Further, I develop the initial framework used as part of the historical analysis section of this thesis, by integrating the case of China into existing historical taxonomies of innovation epochs, using an innovation system perspective.

In the second part, the theoretical concepts underlying the emergence of innovation in China's Strategic Emerging Industries today are presented. This includes an overview of literature on global R&D management in China, the institutional perspective, including the economic and political context of indigenous innovation and Strategic Emerging Industries in China today.

3.1 The historical perspective on innovation

Most analysts agree on the importance of the historical perspective on innovation. Due to the complexity and time-consuming nature of innovation, the outcomes are often only visible after a significant period of time. Furthermore, as Lazonick has noted, "the social conditions affecting innovation change over time and vary across productive activities; hence theoretical analysis of the innovative enterprise must be integrated with historical study" (William Lazonick, 2002). The following sections therefore seek to identify those economic, institutional, political and sociocultural factors that have been found to impact the emergence of innovation from a historical perspective.

The literature on factors driving innovation is vast, making it impossible to cover all aspects of this stream of literature. Previous scholars have traditionally focused on demand side factors of innovation, which influence the potential profits that potential innovators can expect. However, in recent years, scholars are increasingly also considering the supply side factors of innovation, e.g. by considering the role of entrepreneurs as a source of new scientific and technological knowledge (Nicholas et al., 2011). The following section briefly outlines a number of relevant – albeit certainly not all - demand and supply side factors. In the case of demand-side factors, they include the role of political, legal and financial institutions; national innovation systems; intellectual property rights; patents; expected profits of innovation; natural resources; and market size. Supply-side factors include the human capital aspects of entrepreneurship; education and training; culture, religion and language; as well as immigration in the context of entrepreneurship.

3.1.1   Economic context

As mentioned in the introduction, one important reason that has made China an attractive location for R&D and market with significant opportunities arising from innovation is its large *market size*. As previous scholars have noted, the size of the potential consumer market increases the likelihood of innovation, due to increased incentives for innovation based on larger potential future revenues (Acemoglu & Linn, 2004). In the case of China, this creates enormous opportunities for potential innovators and it can thus be seen as an important driver of innovation. While demand factors such as the quality of institutions increase the expected payoff from innovation, the size of the potential market for innovation as well as other physical variables such as natural resource endowment also matter.

Another economic factor that has been found to be conducive to entrepreneurship and innovation from a historical perspective is *factor endowment*. Hicks (1932) theorized that innovators were sensitive to the relative supply of factors of production in their local economy. Fellner (1961) highlighted the existence of market forces directing the factor-saving impact of inventive activity. Habakkuk (1962) suggested that the direction of innovation in different countries is related to the relative scarcity of factors of production in those countries. Davidson (1979) builds on the Heckscher-Ohlin model of comparative advantage theory, which suggests that nations export those commodities in which they possess relatively abundant factors of production. He argues that innovative activities will be concentrated in industries, which intensively use a nation's relatively expensive factors of production. Besides natural resources and physical capital, more recently, scholars have pointed out to the important role of human capital as important pillars of innovation (e.g. Altenburg, Schmitz, & Stamm, 2008).

3.1.2   Institutional & political context

Placing institutions as well as a broader economic, political and sociocultural context at the center of this study's analytical framework, this study seeks to build on the contributions of earlier social scientists such as Thorstein Veblen, who viewed national institutions and their effects on innovation and technological development as a key to understanding differences in economic performance between countries (Murmann & Homburg, 2001). Even though institutions have been neglected in recent contributions on innovation and R&D, and economic analysis in general, this dissertation seeks to move contextual research again onto center stage.

Indeed, institutional quality has been identified as one of the most important foundations of entrepreneurship, innovation and, as a result, of economic growth. Douglass North established institutions as "the rules of the game in a society or, more formally...the humanly devised constraints that shape human action" (North, 1990, p. 1), which "consist of both informal constraints (sanctions, taboos, customs, traditions, and codes of conduct), and formal rules (constitutions, laws, property rights)" (North, 1990, p. 97). In particular, the literature on institutions has focused on property right systems, which have been seen as a positive force for innovation and as a result for economic performance, as they provide incentives and opportunities for entrepreneurs to reap the benefits of their activities (Helpman, 1992; North & Wallis, 1994). In addition, property rights decrease transaction costs, which positively impacts the likelihood of innovation, as they increase potential gains from exchange, therefore increasing productivity and growth (Jones, 2013). For example, Chandler (Chandler, 1992), Freeman (1995a) and Murmann (2000) have studied the emergence of the synthetic dye industry in the second half of the 19th century, considering the importance of the (e.g. national) institutional, economic, political and social context that allowed for the emergence and growth of this industry.

In relation to innovation, scholars have particularly debated the role of intellectual property rights, considering for instance the extent to which patent systems stimulate innovation, the potentially negative effects of patents, and the deadweight losses that arise from monopoly pricing. Patents have been in place since at least 1474, when the Republic of Venice promulgated a decree stating that a new invention could be protected from imitation so long as it was useful, novel, and a working device. These criteria bear a striking resemblance to the ideas underpinning modern patent laws (Nicholas et al., 2011, p. 789). Although formal intellectual property rights diffused across countries over time, some nations were relatively late in adopting patent protection. The figure below shows that more than half of the twenty-nine largest independent countries by gross domestic product (GDP) at the end of the 20th century had patent systems in place by 1850.

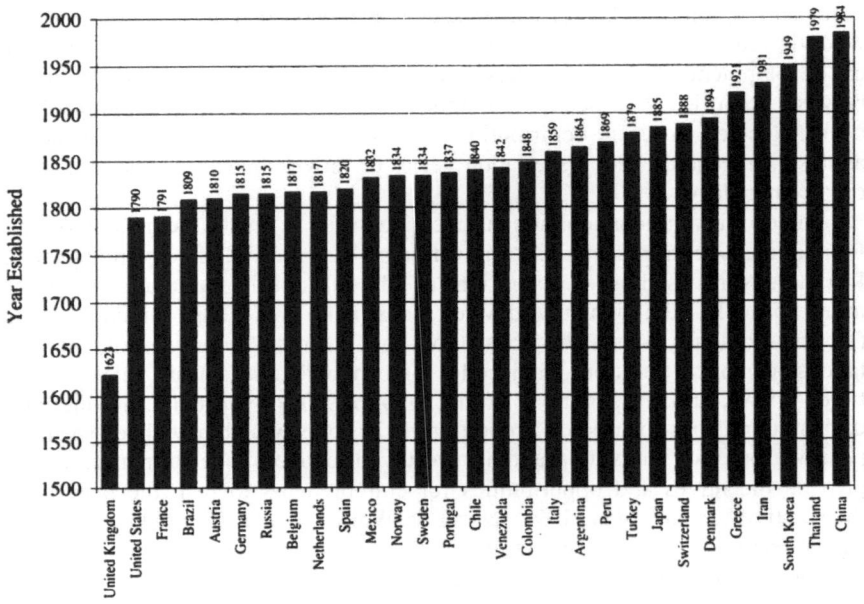

Figure 8: Year of patent system established (Lerner, 2000)

Related to institutional quality, the quality of legal and in particular (intellectual) property rights systems has been identified as positively correlated to the levels of innovation which in turn promotes economic growth (Gould & Gruben, 1996; Helpman, 1992; Lai, 1998). In this context, a great number of studies have examined the effect of patent protection on innovation, with most contributions confirming a positive correlation between IP protection and innovation (e.g. Grossman & Helpman, 1993; Lerner, 2000; Mansfield, 1986). In examining the factors driving higher institutional quality (which in turn positively affects innovation), scholars have pointed to geographical (climate, disease ecology, and distance from the coast) and institutional inheritance through colonialism as possible explanations for existing institutional setups (Acemoglu, Johnson, & Robinson, 2000; Sachs, 2003).

The quality of financial institutions has also been found to contribute to the emergence of innovation. As Neal (1994) and others have shown, the availability of financial institutions – in the form of banks, insurances and capital markets – was important in supplying entrepreneurs with credit to finance large-scale investments, e.g. railroads, steam power and iron works, which laid the foundation for Britain's Industrial Revolution in the 18th and 19th century. Furthermore scholars have noted that in cases where formal institutions are missing or lacking, informal financial networks within particular communities may serve as substitutes to formal financial institutions, as Wolcott illustrates for the Indian context (Wolcott, 2010).

Role of the state

The discussion of what drives innovation must also consider the role of the state and government-related institutions such as the military, all of which have been engaged in promoting innovation in several instances. For example, in the late 19th century, the Prussian government (later imperial Germany) set up technical training institutes (called "Gewerbeinstitute") that served to reverse-engineer British machine tools and disseminate new technology into German industry. This technology transfer was very effective and enabled the German engineering industries to design and manufacture steam locomotives in the 1840s and 1850s (Beer, 1959). The political system has been found to influence not only the amount, but also the kind of entrepreneurship. As Baumol (2002) has argued, liberal market-oriented societies will not necessarily result in more entrepreneurs potentially engaging in innovation. However, entrepreneurs in rent-seeking as compared to liberal market-oriented societies will engage in different kinds of entrepreneurship and are more likely to engage in ventures creating income by wealth re-distribution rather than wealth creation, e.g. through exclusions and special privileges, lawsuits or the manipulation of the political machinery (Joel Mokyr, 2010, p. 3).

More recent contributions have studied National Systems of Innovation – also referred to as NSI – that describe the exchange of knowledge and technology among enterprises, institutions and people in a national context (in particular Freeman, 1995a; Lundvall, 1992; Nelson, 1993). For example, the historical establishment of formalized research and development (R&D) structures within companies and later on in government-run laboratories can be seen as one aspect of such a systemic perspective on innovation.

3.1.3   Sociocultural context

Education and human capital

While the literature generally agrees on the positive relationship between education and economic growth, the link between education, training and innovation is less clear. Intuitively, one may assume that low levels of literacy, education and training will decrease the supply of entrepreneurs capable of conducting innovation. As the stock of human capital is more difficult to increase than the physical stock in a community or country, populations with a high stock of human capital are likely to outperform those with lower human capital stocks in the long-term, in terms of their ability to generate entrepreneurial talent that drives innovation (Sandberg, 1979).

Another area of research has been the role of immigration and in particular that of the immigrant entrepreneur. While foreign inventors accounted for 13 percent of all patents granted in the United States at the beginning of the 20th century, that level had doubled by the year 1970 and further increased to more than half of patents granted in 2009 (Nicholas et al., 2011, p. 800). Other studies have highlighted the role of immigrant communities and informal networks in driving entrepreneurship, for example in the context of ethnic Chinese networks in international trade (Rauch & Trindade, 2002).

The role of (national) culture

A more contested potential driver of innovation is culture. A great number of scholars have engaged in research that links culture to the level of entrepreneurship and innovation in a given community in order to explain why some regions have seen lower levels of entrepreneurship and innovation compared to others in the past, also explaining their lower levels of economic development.

A number of scholars follow Hofstede (1980) who defined culture along four key dimensions: power distance (the degree to which the less powerful members of a society accept and expect that power is distributed unequally); uncertainty avoidance (the degree to which the members of a society feel uncomfortable with uncertainty and ambiguity); individualism (the degree of preference for a loosely-knit social framework in which individuals are expected to take care of only themselves and their immediate families); and finally masculinity versus femininity (the preference in society for achievement, heroism, assertiveness, material rewards for success and competitiveness, as compared to cooperation, modesty, caring for the weak, quality of life and consensus-orientation).

In 1993, a fifth dimension ("long-term orientation") was added, which takes into account the general principles of the Confucian value system, which - according to the study - is based on future orientation, thrift and persistence (Hofstede, 1993). While Hofstede has made important contributions for the *measurement* of different dimensions of culture, several critiques of his work have pointed out the weaknesses of his theory in *comparing* the resulting scores across cultures. Other possible limitations include an oversimplification of cultural differences, inconsistencies between his categories, lack of empirical evidence from educa- tional settings and overall a model of culture as static (instead of dynamic) (Signorini, Wiesemes, & Murphy, 2009).

Another cultural dimension that scholars have argued to be influencing the level of innovation is the notion of a protestant work ethic (Weber, 1904), a concept that has been critically assessed by a number of scholars (e.g. Landes, 2000; Van Hoorn & Maseland, 2013). Furthermore, the relationship between religion and values (e.g. Confucianism) on the one hand, and entrepreneurship on the other hand has been studied, with most studies confirming a significant correlation (e.g. Audretsch, Boente, & Tamvada, 2007; Hofstede & Bond, 1988). Similarly, language has been identified as a potentially influencing innovation, by enabling or disabling access to innovation networks and potential markets.

While the context factors outlined above are but an imperfect list of all aspects affecting innovation in a historical context, they provide a starting point for the analysis of innovation from a historical and contemporary perspective, and how the relative importance of factors may have shifted over time.

In conclusion, in reframing contemporary management issues with a view to its long-term meaning, the historical perspective offers a multi-disciplinary approach, which can illuminate government-business relations, technology, corporate culture and business ethics (Knowles, 2004). Modern managers operating in a business environment marked by continuous and rapid change need to be aware of how long-term changes have affected enterprises in the past, and what we can learn from these experiences. This can also help them make more informed decisions based on a long-term perspective.

3.2 Innovation in China's Strategic Emerging Industries

Most scholars agree that country-based contexts matter for business operations and that the beliefs, goals and actions of firms and individuals are strongly influenced by environmental and institutional settings (most notably Boisot & Child, 1996; Bruton & Ahlstrom, 2003; Child & Tse, 2001; Clarke, 1991; North, 1990; Scott, 1987, 1995) in subtle but pervasive ways (Bruton & Ahlstrom, 2003, p. 234). In particular, "managers and firms rationally pursue their interests and make strategic choices within the formal and informal constraints in a given institutional framework" (Peng et al., 2009).

Therefore, the institutions-based context of China is likely to differ significantly from the one encountered in traditional innovation locations such as Western Europe and the US. China's current institutional framework may be considered as still relatively weak as compared to developed countries from a western perspective. While research on the influence of institutional context on organizational and commercial systems is limited in general (Mahoney & Chi, 2001; Peng & Luo, 2000), this is especially true in the area of entrepreneurship and innovation (Giamartino, McDougall, & Bird, 1993). The institutional and political aspects of innovation in China will be further developed in 3.2.2. The following section outlines the relevant theory on the economic context of China.

3.2.1   Economic context of China

As mentioned, the size, structure and competition landscape of markets are important determinants of innovative activity. In China, as the country is still undergoing a fundamental transition from a centrally planned to a market-oriented economy, firms have to conduct business and innovation in an incomplete market environment (Peng, 2002). In recent years, firms in China's domestic market have steadily moved up the value chain, from engaging in standardized manufacturing activities towards increasing technological sophistication, with levels of innovative capabilities increasing (Linden et al., 2009). Partly, this drive towards more innovation is due to macro-economic developments, such as increasing wage levels in coastal as well as inland regions of China, which provide an incentive for foreign as well as local companies to move into higher value-added manufacturing activities. On the other hand, China's rapid industrialization process has had significant side effects, most notably on its environment, e.g. in the form of air and water pollution, severe traffic congestion especially in urban areas, as well as water shortages. In this context, innovation in the area of renewable energy products, as for instance in solar, wind or thermal energy, as well as in the development of battery-powered automobiles have attained increased attention in China.

The large size of the Chinese market is conducive to standardization and mass production of new innovations. These conditions are conducive to the establishment of good-enough, widely available products (e.g. Widenmayer, 2012; Zeschky et al., 2011; Zeschky, Widenmayer, & Gassmann, 2014). These characteristics suggest that China's market may be conducive for disruptive innovation rather than strictly technological innovation, in which China is currently catching up with established centers of innovation.

For firms conducting innovation in China, the country's ongoing transition context has several important consequences. On the one hand, Chinese policymakers have actively supported the economic transformation of the economy, moving from a centrally planned to a market-based structure. In this transition, firms are often influenced by institutional pressures aimed at inducing foreign and domestic firms to conduct innovation locally (Tan & Tan, 2005). In China, besides a competitive market environment, the government has been identified as a significant external driver of technology-related innovation (Yang et al., 2010).

On the other hand, as changes in government innovation policies have important repercussions, companies adopt revised innovation policies to adjust to China's particular context. Hence, previous studies have found that the uncertain environment of China's transition economy strongly influences firms' technological innovation, due for instance to the uncertain and competitive market environment (Li, Liu, & Zhao, 2006).

In comparing firms conducting innovation in developed economies as compared to China, prior contributions have further shown that "during the transition from a planned to a market economy, firms in China have preferred a more defense-oriented strategy when environmental uncertainty is high." (Yang et al., 2010, p. 823). Furthermore, one study concludes that higher environmental uncertainty negatively impacts the levels of proactiveness, future orientation and risk-taking in innovation strategies (Zhou, Yim, & Tse, 2005). A comprehensive literature review of 175 articles on technological innovation of firms in the context of China from 2012 concludes that "different results from existing studies challenge us to inquire into the following questions: Do these environmental factors have different influences on different kinds of innovation? How do these environmental factors affect the technological innovation of different firms in China?" (Yang et al., 2010, p. 823). Closely related to the context of China's domestic market context are government policies, which continue to shape the country's innovation environment.

3.2.2 Institutional & political context

Emerging markets like China are also marked by rapid change (Peng, Wang, & Jiang, 2008), making them a useful setting for understanding the impact of institutions on a given industry undergoing a rapidly changing environment (Bruton et al., 2009). Therefore, it is essential to understand China's institutional and political context in order to evaluate innovation in its Strategic Emerging Industries (SEI). This section provides a theoretical outline of institutional perspectives on management, as well as the political economy surrounding China's SEI.

Institutional perspectives on management in China

According to Institutional Theory, the beliefs, goals and actions of firms and individuals are strongly influenced by environmental and institutional settings (Bruton & Ahlstrom, 2003; R. Scott, 1995; W. R. Scott, 1987). There are two broad approaches to conceptualizing Institutional Theory: sociological and economic. They are depicted in the figure below.

Institutional Theory (IT)

❶ Sociology & Organization Theory View (DiMaggio & Powell, 1991)	❷ Economics & Political Science View (North, 1990; Shepsle, 1989)
Principle driving forces	
▪ **General:** Effort to achieve legitimacy & stability in uncertain situations; values, views and norms of **social classes** (Zucker, 1991)	▪ **General:** Governance structures and rule systems constructed by individuals; "rules of the game" (North, 1991)
▪ **Human behavior:** Social norms, shared cultures, cognitive scripts (Meyer & Rowan, 1991)	▪ **Human behavior:** Rules and procedures, formal control (North, 1991)
▪ **Relationship: Institutions–Organizations:** Organizations adjust and conform to values and limits prescribed by a society's institutions	▪ **Relationship: Institutions–Organizations:** External institutions create structures for organizations (North, 1991)

Figure 9: Perspectives on Institutional Theory

The sociological perspective on Institutional Theory emphasizes the legitimacy-building and role-shaping actions of institutions (Suchman, 1995), as well as behaviors that arise from shared cultural and political systems (W. R. Scott, 1987; Zucker, 1987). According to this view, organizations and commercial practices exist due to the taken-for-granted nature of institutions and their self-sustaining ability (Bruton & Ahlstrom, 2003; DiMaggio & Powell, 1991).

In contrast, the economic perspective on Institutional Theory largely draws on the neoclassical economics view of Douglass North (1990), arguing that the institutional framework of a society provides a formal rule framework regulating economic activities, which he labels as "the rules of the game" (North, 1990). Accordingly, the relevant framework is a combination of political, social, and legal rules that defines a basis for production, exchange, and distribution in a system or society (North, 1990; Roy, 1999).

In order to operationalize the concept of Institutional Theory for the purpose of this dissertation, we follow Scott (1995) in categorizing these formal and informal institutions into three groups: normative, regulatory, and cognitive institutions. The first set of institutions, regulatory institutions, are usually defined by the legal system and are the most formal institutions. Normative institutions are more informal less codified and determine the roles and actions of the individual. Examples include generally accepted authorities, for instance technical or medical associations. The third group, cognitive institutions, includes the least formal and includes social norms and behaviors, which are taken-for-granted or emerging through social interaction in communities or countries (Bruton & Ahlstrom, 2003). In accounting for the impact of different institutional factors in the management of innovation, the study focuses on these three categories of institutions. This framework is shown in the following illustration.

Types of Institutions
(Scott, 1995)

1 REGULATORY	**2** NORMATIVE	**3** COGNITIVE
• Government legislation and industrial agreements/standards (Bruton et al., 2010) • Guide behavior **through rules of the game,** monitoring & enforcement (North, 2000)	• Models of organizational and individual **behavior** • **Based on values and norms** that create ground rules to which people conform (Scott, 1995)	• Models of **individual behavior** based on **subjective** or constructed rules & meanings • **Examples:** culture; language; "taken-for-grantedness", societal acceptance of entrepreneurship

Figure 10: Types of institutions

Building on the different categories of Institutional Theory as outlined above, China's institutional environment differs from the Western context in all of the three dimensions (Bruton & Ahlstrom, 2003; He, Tian, & Chen, 2007; Peng & Luo, 2000). For instance, as typical of many emerging economies, China's regulatory and legal enforcement regime is still inadequate when compared to Western standards. Due to inadequately enforced laws to enforce contracts and claims such as intellectual property rights, companies in China often need to seek alternative ways of conducting business, e.g. through less formal mechanisms in the form of personal relationships, adapted business processes or private security arrangements (Bruton & Ahlstrom, 2003). In the context of innovation in China, the issue of intellectual property protection in China lies at the heart of discussions about innovation in China. The following section therefore discusses this important aspect of China's institutional framework in more detail.

Protection of Intellectual Property Rights (IPR)

As mentioned, as typical of many transition economies, China's regulatory and legal enforcement regime is still less developed compared to Western standards (Boisot & Child, 1996; Peng & Luo, 2000). Before China's economic opening period starting in 1978 under Deng Xiaoping, China had no intellectual property rights (IPR) protection. The first patent law was established in 1985 and substantially reformed in 1992 and 2000 (Awokuse & Yin, 2010). Patent laws were further strengthened in the wake of China's membership in the WTO in 2001, which required Chinese IPR laws (i.e. patents, copyrights and trademarks) to be more aligned with the WTO's Agreement on Trade-Related Aspects of Intellectual Property Rights (TRIPS) and other international IPR conventions (Awokuse & Yin, 2010, p. 218). Foreign as well as domestic patents in China have significantly increased since 1992. The majority of Chinese patents (invention patent applications) were filed by foreign firms, with most patentees coming from Japan, the US, EU member countries, and South Korea, as shown in the illustration below.

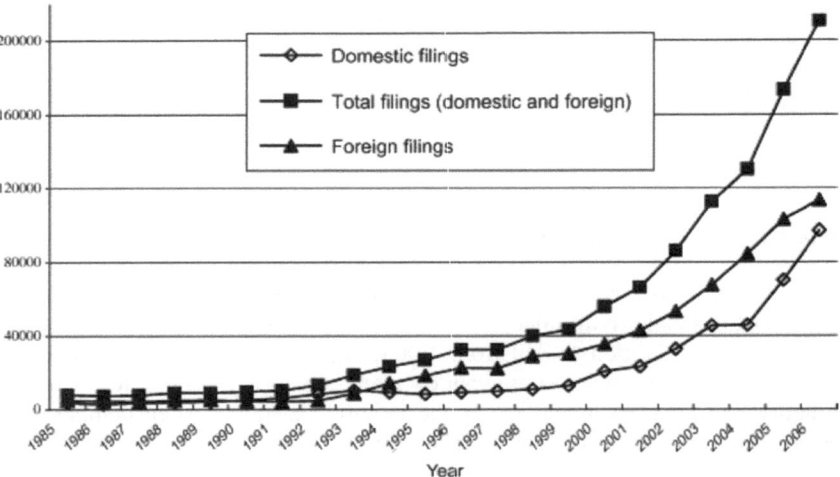

Figure 11: Patent filings in China from 1985-2006 (SIPO, 2014)

Nevertheless, despite the increasing number of patents registered in China, many firms innovating in China suffer from challenges related to the insufficient legal enforcement of IPR protection in China. In recent years, the scale of intellectual property theft has dramatically increased in terms of volume, range of goods, countries affected and sophistication (Gassmann et al., 2012, p. 17). A survey of the German Engineering Federation conducted in 2012 showed that 67 percent of respondents (from manufacturing and industrial goods firms) claimed to have suffered from product piracy (Rotter, 2013). Furthermore, the protection of firm intellectual property in China is likely to remain weak as long as domestic firms' demand for tighter control remains relatively low (Fuchs, Kammerer, Ma, & Rehn, 2006).

Therefore, due to inadequately enforced laws to secure contracts and to make claims based on intellectual property rights, foreign companies in China often need to seek alternative ways of conducting innovation, e.g. through less formal mechanisms in the form of personal relationships, adapted business processes or private security arrangements (Bruton et al., 2009). These formal and informal networks between decision makers in the Chinese central and local governments on the one hand, and Chinese and foreign companies on the other hand, can provide the latter with market information and other strategic considerations (Li, Poppo, & Zhou, 2008). Due to this particular institutional environment of China, it is unlikely that industry players in China operate under similar or even equivalent premises as European, American and even Japanese or Korean companies. Instead, as suggested by previous authors in the context of other industries, it is reasonable to assume that institutional factors in China may be lead to industries with particular, idiosyncratic characteristics (Bruton & Ahlstrom, 2003). In particular, this may also be true for the way in which companies generate innovation in China as compared to developed markets such as Europe.

The following section outlines Chinese government policies targeted at increasing the domestic level of innovation. In particular, this dissertation focuses on measures to increase the level of innovation in China's Strategic Emerging Industries.

Chinese government policies in Strategic Emerging Industries (SEI)

In China, the role of the government still pervades firms in their daily operations, as well as in their long-term strategic planning. In particular, it influences firms' innovation processes with innovation-related policies, institutional arrangements and provisions (e.g. on how foreign companies may operate and innovate in China), as well as through direct intervention, e.g. in industries of public interest such as the steel industry (Huang, Schroder, & Steffens, 1999). Especially due to the environmental and demographic challenges that China is facing, both its central and province level governments have made it a strategic priority to incentivize domestic and foreign companies to develop more innovative products.

At the annual gathering of the National People's Congress on Wednesday, 5 March 2014, China's Premier Li Keqiang announced that "we will declare war on pollution and fight it with the same determination we battled poverty" and further that pollution "is nature's red-light warning against the model of inefficient and blind development" (BBC News, 2014). Indeed, since the mid-2000s, China has witnessed a paradigm shift of technological innovation, characterized by a policy change from focusing on expected technological spill-over from multinational corporations to emphasizing indigenous innovation conducted by domestic as well as foreign companies. The following section describes in more detail the measures that Chinese policymakers have taken in the last years to increase the level of innovation in China.

Government policies to promote SEI

While China has been perceived as a strategic location for low- and medium-value manufacturing, increasing wage levels in the manufacturing sector have reduced some of the cost advantages of moving production to China. Largely based on these developments, as well as the experience of the recent global financial crisis, the Chinese government has made it one of its top priorities to move the domestic economy further up the value chain. For this reason, a great number of laws, policies, guidelines and regulations have been introduced to support the development of the Chinese economy increasingly towards the development and production of more sophisticated, higher-value products, with higher levels of technological innovation. Examples include regulations on technology transfer for foreign investments and joint ventures in China, as well as government procurement to support domestic SEI.

After the financial crisis of 2007-08, the Chinese government was seeking new ways to boost long-term economic growth and international competitiveness, to restructure and diversify the economy away from its previous manufacturing-led growth in order to increase resistance to economic shocks, and to make China into a globally competitive innovative economy. In this context, the SEI initiative can be seen as a core part of China's overarching state-supported strategy to stimulate innovation, technological leapfrogging and economic catch-up with developed economies and also address increasing socio-economic and environmental challenges in China (Wen, 2009, November; Wen, 2013, February).

These efforts are aimed at narrowing China's technology gap with world leaders to increase competitiveness, so as to avoid the difficulties that other (e.g. Latin American) countries have experienced in moving from a middle-income to a high-income status, a phenomenon that has become known as the middle-income trap. According to the 2006 Medium and Long-term Science & Technology Plan China is determined to become an "innovation-oriented" country by 2020 and a "leading science power" by 2050, as well as to reduce its dependence on technology from other countries to 30 percent or less (2006 State Council, 2006). An important component of this drive is the development of "indigenous innovation" (called "zizhu chuangxin" in Mandarin), which encourages domestic innovation through "original innovation", "integrated innovation" – i.e. combining existing technologies in a new way - and "assimilated innovation" – i.e. making improvements to imported technologies (2006 State Council, 2006).

In September of 2009, China's then prime minister, Wen Jiabao chaired a meeting that helped identify the seven major SEI industries that should support Chinese long-term economic development, and on October 10, 2010, the "decision about accelerating the cultivation and development of strategic emerging industries" was formally promulgated by the Chinese State Council, opening the door for substantive support of SEI in China in the following years. In 2012, when China was again facing a global economic slowdown, Zhang Xiaoqiang of the Chinese National Development and Reform Commission declared that "when the economic outlook is not good, developing strategic industries will definitely help alleviate downward pressure on the economy" and further that "we have to take a longer view and develop new bright spots of growth" ("China eyes new strategic industries to spur economy," 2012).

China's strategic emerging industries (called "zhanlue xing xinxing chanye" in Mandarin) are an important part of China's recent indigenous innovation strategy. SEIs are a group of seven industries and related sub-industries, the development of which is a core part of China's state-led strategy to boost economic competitiveness and to ensure sustained long-term growth after the global financial crisis of 2007/8. More specifically, the SEI initiative is intended to stimulate rapid technological catch-up and ultimately economic convergence with developed economies, while also addressing increasing socio-economic and environmental challenges in China. After consulting with stakeholders on the initiative through a series of seminars in 2009 organized by then Premier Wen Jiabao and Vice Premier Li Keqiang, the SEIs initiative was officially promulgated in the "Decision on Accelerating Development of Strategic Emerging Industries", issued on 10 October, 2010 (State Council, 2010).

The concept of "strategic emerging industries" (SEIs) and the policy initiatives for their development, the first of which were officially promulgated in 2010, form the foundation of what the Chinese government hopes will "leapfrog" their country to the forefront of the world economy in the coming decade, while also addressing socio-economic and environmental challenges at home. The envisioned amount of investment in SEI – 10 trillion RMB from 2011-15 – suggests that the Chinese government is clearly serious about developing SEI (USCBC, 2013). The particular SEIs, as well as their sub-industries, are provided in the table below.

Table 3: The seven SEI and their sub-industries

Industry	Sub-industries
Energy conservation and environmental protection	Key technologies, equipment and products for efficient energy conservation; critical generic technology R&D for resource recycling; advanced environmental protection equipment and products; market-based energy conservation and environmental service system; waste and old product recycling and reuse system underpinned by advanced technology; clean coal and seawater utilization
New-generation IT	Information network infrastructure, new generation mobile communication; new-generation Internet; new generation of core equipment and intelligent terminals; ; three-network convergence; Internet of things; cloud computing; IC; new displays; high-end software; high-end servers; software services; internet-based value-added services and other information services; intelligent infrastructure; digital virtualization
Biological/Biotech	Biotech medication, new vaccines and diagnostic agents, chemicals, modern Chinese medicine and other innovative medicine; biopharmaceuticals; medical devices, materials and other biomedical engineering products; bio-breeding; green bio-products for agricultural use; bio-agriculture; marine-origin biological technologies/products; bio-manufacturing
High-end equipment manufacturing	Main and branch line aircrafts, generic aircraft and other aviation equipment; space infrastructure; satellites and application thereof; passenger special lines, urban railway transportation and other rail traffic equipment; marine engineering equipment; intelligent manufacturing equipment based on digital, flexible and system integration technologies
New energy	New-generation nuclear energy and advanced reactors; solar energy utilization; photovoltaic and photo-thermal power generation; wind power technological equipment; intelligent power grids; biomass energy

Industry	Sub-industries
New materials	Rare earth functional materials, high-performance membrane materials, special glass, functional ceramics, semiconductor luminous materials and other new functional materials; high-quality special steel, new types of alloys, engineering plastics and other advanced structural materials; carbon fibers, aramid fibers, ultrahigh molecular weight polyethylene fibers and other high-performance fibers and compound materials; nanometer, superconductive, intelligent materials and other common basic materials
New energy automobiles	Power batteries; pure electric vehicles; driving motors; electronic control; plug-in hybrid electric vehicles; fuel cell automobile technologies

Source: Chapter 10, 12th Five-Year Plan (2011-2015) for the National Economic and Social Development of PR China (China Central Government, 2011)

Besides supporting China's long-term economic development more broadly, Chinese policymakers also view SEI policies and measures as essential in addressing the most commonly identified barriers to the development of innovative, future-oriented industries in China in general. These include a lack of indigenous innovation capacity, core intellectual property (IP) and technology; a changing IP landscape led by incumbents and resulting in increased licensing fees and costs of R&D; a low technological achievements conversion rate; a shortage in creative talent and qualified human resources; and a sub-optimal financing system (Prud'homme, 2015, p. 3).

Typically, the central government sets targets to support the growth of Strategic Emerging Industries, which are subsequently implemented and administered at the provincial or city level. For example, local governments would typically provide funds available to innovation-related projects, often with a focus on building up SEI-based clusters in a particular region, e.g. focusing on the development of the energy sector in the Shanxi province.

The following section provides a brief overview of the key agencies – at the central and local levels - involved in driving the development of SEI in China, followed by an outline of selected government policies supporting SEI. In addition, the subsequent section introduces a list of state-created intellectual property (IP)-related measures (labeled as "SIPMs" hereafter) related to the creation, management, utilization, and protection of IP. These include Chinese policies, laws, regulations, guidelines and other measures and practices of central and/or sub-central level governments and/or state-affiliated entities, which from the perspective of European companies may turn into barriers (state-created intellectual property (IP)-related barriers, or "SIPBS" hereafter) and which are at the center of this study.

Key stakeholders in the promotion of SEI in China

Similar to other broader policy plans in China, the creation and implementation of SEI related policies follow a similar pattern, in which the central government provides overarching policy guidelines and in which implementation is done at the provincial or community level (USCBC, 2013, p. 3). For companies, it is essential to understand both the overarching vision of the Chinese central government for future industry development including SEI, as well as the perspective of local governments and institutions that seek to implement them, often incentivized by innovation-related key performance indicators (KPI) such as number of patents or local R&D facilities.

Several agencies are involved in formulating, coordinating, implementing and monitoring SEI policies. The 12th Five-Year Plan initiated the formation of an inter-ministerial coordination group, which is led by the National Development and Reform Commission (NDRC) (China Central Government, 2011). This group further includes the ministries of Commerce (MOFCOM), Science and Technology (MOST), and Industry and Information Technology (MIIT). This group coordinates, analyzes, and tracks the implementation of SEI policies across relevant agencies. Some of these agencies also have the authority to approve projects at the municipal, provincial, and national level (USCBC, 2013, p.3).

Following this overview of the institutional and political context of innovation in China's SEI, the next section focuses provides a better understanding about the sociocultural context of innovation in China.

3.2.3   Sociocultural context

As much as successful company and government policies matter, it is ultimately people that drive innovation. Individuals are shaped by their communities, as well as their local context. As various regions and nations have produced different levels of innovation over time, a number of scholars have investigated the relationship between (national) sociocultural context and national innovation rates. In particular, the relationship between culture and innovation level has been studied. Culture refers to a core set of attitudes and practices that are shared by the members of a collective entity, such as a nation or a firm (Hofstede, 2003). It reflects a country's "central tendencies" in terms of values, beliefs, and preferences (Hofstede, 1991). This interpretation generally matches the definitions of culture used by culture and innovation scholars across several fields (Taylor & Wilson, 2012, p. 236).

In relation to innovation, scholars have focused in particular on cultural values such as individualism, power distance, uncertainty avoidance, and masculinity in relation to national rates of innovation – typically measures by the number of trademarks or patents per capita. In particular, the level of uncertainty avoidance, as well as power distance and individualism are correlated with innovation levels. Most significantly, several studies have found that most measures of individualism have a strong, significant, and positive effect on innovation, even when controlling for major policy variables (Taylor & Wilson, 2012, p. 241). However, Taylor and Wilson (2012) have also shown that a certain type of collectivism - patriotism and nationalism - can also foster innovation at the national level. In contrast, other types of collectivism (i.e. familism and localism) not only harm innovation rates, but may hurt progress in science worse than technology (Taylor & Wilson, 2012, p. 240). While actual national levels of patriotism and nationalism are difficult to measure, media in recent years have pointed out to Chinese policymakers increasingly encouraging patriotism among Chinese citizens (Foreign Policy, 2014). The following section on China's national culture provides further insights into underlying factors that influence the emergence of innovation.

China's national culture

China's national culture is another element of the country's innovation environment that has been found to influence companies in generating innovation. In this context, it is very important to note that China is a multicultural country made up of 55 ethnic groups, in addition to the Han majority of the population, which represent different cultural values, religions, and make up about 8.5 percent of the national population ("Han Chinese proportion in China's population drops: census data," 2011), which translates to almost 115 million ethnic minorities in China's large population. The following description of China's national culture can therefore merely be seen as a rough approximation that cannot take into account the complexity of China's national culture, while still providing valuable insights for the purpose of this dissertation.

Previous extensive research suggests that there are three aspects of national culture that are likely to influence innovation: a nation's religion, its geographic location, and the values of its citizens (Hofstede, 2001). Dimensions along which cultures differ include individualism–collectivism, uncertainty avoidance, power distance, masculinity–femininity, and long-term orientation. More recent contributions have shown which impact these dimensions have on innovation (Dwyer, Mesak, & Hsu, 2005).

For Asian countries such as China, previous studies have shown that national culture and administrative heritage in Pacific Asian countries have a significant impact on the way that knowledge transfer and innovation occurs (Lu, Tsang, & Peng, 2008). For example, Flynn (1985) found that in the case of Japan, its culture and management style emphasizes consensus building, positively influencing the ability to engage in innovation involving modification, improvement, and the application of technology, but negatively related to the ability to invent or discover revolutionary new technology (Flynn, 1985, p. 159). Similarly, in China as another society based on Confucian principles, harmony, cooperation, and convergence are traditionally valued highly (van Someren & van Someren-Wang, 2013, p. 31). This means that when evaluating the innovation performance of different regions, it is important to keep in mind the local sociocultural context, which determines not only the level of innovation, but also the kind of innovation that is more likely to occur.

Furthermore, while Western countries often have low context cultures (e.g. Germany and the United States), China, like many other Asian countries, is a very high context culture (Hofstede, 1991). This means that Western managers tend to convey messages mostly by words, and believe that formulating ideas clearly enough is sufficient to avoid ambiguity. In contrast, in high context cultures like China, messages are delivered through nonverbal signals (e.g. through the tone of voice, facial cues, use of silence, and body language), unspoken assumptions, and the context or environment of the conversation. Employees from high context cultures such as China may assume that the receiver of the message is able to understand the true connotation of a message. These differences in culture and communication styles may therefore affect the interaction between foreign (e.g. German or American) and Chinese R&D engineers and potentially lower the effectiveness of cooperation (Gassmann et al., 2012, pp. 8–9).

Organizational culture in China

The specific cultural aspects of China as a country, some of which were outlined above, also affect culture at the industry and organizational level. In recent years, a growing body of literature has sought to describe how national culture affects organizational culture. The latter can be defined as a system of shared assumptions, beliefs, and values, which affect the way in which people behave in organizations, e.g. how they interact, and how they perform in their jobs.

Organizational culture is unique to an organization and can vary along several dimensions.

For instance, organizations can have *strong or weak cultures*, depending on the degree to which its members share and commit to its core values. The higher the sharedness and commitment, the stronger the culture increases the possibility of behavior being consistent amongst its members. In contrast, a weak culture opens can increase the importance of individual member concerns.

Furthermore, organizational culture can be distinguished in terms of *formal and informal culture*. Roles, responsibilities, regulations, accountability and rules are components of formal culture. They set the expectations that the organization has from every member and indicates the consequences if these expectations are not fulfilled. In contrast, informal culture has tangible and intangible, specific and non-specific manifestations of shared values, beliefs, and assumptions, e.g. through symbols, rituals, or shared stories.

Finally, especially for the case of China, the distinction between *soft and hard culture* is highly relevant. Soft cultures typically emerge in organizations that pursue multiple and sometimes conflicting goals. *Soft culture approaches* seek *to* influence employee behavior by *nurturing* people to build a commitment to doing a good job, rather than by pressuring people to do things they would not have freely chosen. Soft and hard corporate cultures have also been described as authoritarian or autocratic as compared to democratic styles. Both approaches can be found in Chinese organizational culture: "yang" (hard, powerful) is contrasted with "yin" (soft, nurturing); these different views find expression in proverbs such as "spare the rod and spoil the child" as compared to "you catch more bees with honey than vinegar".

In China, employees, which is a highly collectivist society, employees expect autocratic leadership because their value system presupposes the manager – who is usually older - to be more experienced and wise. This is in contrast to Scandinavian countries, where decision-making authority is decentralized and built on consensus. Similarly, in Japan, employees prefer to make a decision by consensus rather than majority. Everybody in the group has to agree on an idea before the group takes action. Japanese style decision-making therefore focuses on understanding multiple alternatives rather than a single "correct" answer. In contrast, Chinese managers tend to be more hierarchical in their decision-making processes. They tend not to ask employees for their ideas, but to make the decisions themselves. Thus, the power distance between employer and employee is significantly larger in China than it is in Japanese (or American and European) managerial systems. Thus, Chinese typically seek to maintain social order through a "harmony-within-hierarchy arrangement". Furthermore, while Japanese corporate and industrial relations are generally based on employee loyalty, docility and sacrifice, in China, loyalty towards employers is less pronounced, as high turnover rates suggest (further information is provided in the case study analysis) ("Culture shock: Japanese firms in China," 2010).

For the development of innovation, these characteristic features of Chinese organizational culture matter: the development of something new involves experimentation, trial-and-error, and willingness to accept ambiguity and uncertainty. China's traditionally seniority as compared to competency-based hierarchical corporate structure has therefore been seen as an important impediment for innovative thinking. Due to the increasing level of internationalization within Chinese companies and the increasing influence of Western management models, this is slowly changing, as the analysis of case studies will further elaborate on.

One important mechanism allowing for innovation to occur in China is *trust* among different stakeholders within the organization. As China's *culture of trust* is also different from Western approaches, the following section provides further insights on this.

Culture of trust: China in international comparison

Trust is usually defined as a positive expectation that the other party will act in honest and benevolent ways, reducing fear that one may be exploited. In China, as in any other country, a trusting relationship is an essential prerequisite to the development of the business process. Establishing relationships on the basis of trust is important to all human beings. In business relations, in most Western cultures, trust is typically the default: business partners may be considered trustworthy – based on shared interests – unless something breaks that trust; trust is being built up along with the business transaction. In China, however, the default is tilted more towards distrust. Therefore, business partners need to invest more upfront in order to establish a relationship of trust in order to be able to engage in business. In the West, trust is used to explore and establish – if possible - fertile ground for future opportunities. In China the primary function of trust is to establish and protect feelings of safety at the beginning. In China's relational culture, tight social networks shape Chinese society: trust exists between people in the same guanxi, but it is never assumed outside of it. Therefore, distrust becomes the default in China and trust can only be established after significant relationship-building (Cremer, 2015).

In relation to innovation, this matters especially when organizations are multicultural: in the R&D department of a European industrial company operating in China, managers need to invest in relationships with their business partners, as well as internally with their subordinates. Failure to do so may lead to sub-optimal results as the level of trust may be inadequate and different perceptions interaction across hierarchies may impede organization-internal knowledge transfer. The following sections outline further cultural aspects of innovation management in China, which are of academic and managerial relevance.

China's culture and the protection of intellectual property

In relating Chinese traditional culture to innovation and issues such as the protection of intellectual property, previous studies have highlighted the cultural origin of differences and lack of IPR awareness in China. Traditionally in Confucian societies, the imitation and reproduction of ideas, scholarship and art is seen as a sign of respect and honor. Copying a master's work is seen as an integral part of the learning process between student and teacher, rather than as part of an individual pursuit (Cheung, 2009, p. 20). In a western context, this may be true to an extent only in an artistic context, but not in a business context. It is important to note that due to the complexity of "culture", concluding remarks about the effects of "western" as compared to "Confucian" cultures on innovation need to be seen as approximations at best. Furthermore, culture is an evolving phenomenon, making conclusive interpretations nearly impossible.

In general, these observations suggest that the cultural context of Confucian societies such as China may be favorable towards innovations based on modification, pragmatic simplification and improvement, all of which potentially contribute to disruptive and/or frugal innovation. Prior contributions on the management of R&D in China have shown that cultural factors indeed influence the way in which employees in R&D departments work (e.g. the need to save fact and to win trust), providing a link between national culture and the emergence of innovation (von Zedtwitz, 2004).

Finally, the importance of 'guanxi' (meaning 'relationships' in Chinese) is another important contextual factor. Park and Luo have shown that the utilization of guanxi as an important relationship element embedded in Chinese culture can lead to higher firm performance. Furthermore, Chinese firms use guanxi as a way to overcome competitive disadvantages caused by China's inadequate legal, financial and political institutional framework by "cooperating and exchanging favors with competitive forces and governmental authorities" (Park & Luo, 2001, p. 455).

Having gained a better understanding of the most relevant aspects of China's institutional and political context that have been found to affect innovation, we will now integrate the historical perspective on factors driving innovation with the situation encountered in contemporary China.

3.3   How history can inform the present: research framework

The main contribution of this thesis is to extend our current understanding of innovation in China's Strategic Emerging Industries by integrating a historical perspective. Using a case study as well as historical approaches, this thesis considers changing patterns in the context of innovation that have occurred in successive time periods, with a focus on economic, political and institutional, as well as sociocultural factors. While current analysis of innovation in China often highlights the singularity of the Chinese experience, this research uses an institutions-based perspective to shed light on both the continuities as well as evolutionary and disruptive aspects of innovation in China today, based on a historical perspective.

To this end, below, an initial reference framework is developed based on the existing literature on R&D management, the institutions-based view on management, as well as the historical approach. The framework allows for the reflection and integration of current theory and literature on the phenomenon under investigation.

The initial reference framework allows the reflection of existing theory and literature on the phenomenon under investigation. It includes the categories, their dimensions, and relationships relevant for addressing the research question as presented in chapter one. Additionally, the initial reference framework guides data collection and data analysis. During the research process, the initial reference framework is reflected based on empirical insights and adapted when necessary. The research will result in a conceptualization of innovation in China's Emerging Industries from an institutional-based and historical perspective. This conceptualization will address the research question. Subsequently, the empirical insights will be reflected on existing theory, leading to propositions that expand current literature (Widenmayer, 2012, p. 102). The following section describes the initial reference framework and its derivation, including its underlying theoretical assumptions based on the relevant literature.

The *historical context* of innovation outlines those economic, institutional and political, as well as sociocultural factors that have been identified in the literature as conducive or deterring innovation, based on a historical research approach. The historical analysis includes the three distinct time periods outlined earlier: the First and Second Industrial Revolutions, as well as the postwar period.

Economic factors that have been identified in the history literature as affecting (national) levels of innovation include market size, as well as factors endowment, e.g. natural resources. *Institutional and political factors* that are examined from a historical perspective include, amongst others, formal and informal institutions, the role of intellectual property rights and their protection, the role of the state in promoting innovation, as well as the concept of National Innovation Systems and their contribution in historical perspective. Regarding *sociocultural factors*, this thesis focuses in particular on the role of education, human capital and (national) culture for innovation, from a historical point of view.

Mirroring the historical perspective on innovation in structure, the *Context of China* describes the contemporary economic, institutional and political, as well as sociocultural context of innovation in China. Using case-based evidence of four leading Chinese and non-Chinese multinational corporations conducting R&D and innovation in China, this part identifies drivers of innovation in the context of contemporary China, which are used later on as a basis for evaluating the context of China from a historical perspective. In terms of *economic factors*, the size and dynamism of the Chinese market, as well as its continuing rapid transformation from a centrally planned to a market-oriented economy and its consequences are discussed. *Institutional and political* are important influencing factors in China's for all firms operating and conducting innovation in China. Focusing on the empirical evidence from four companies operating in Strategic Emerging Industries in China, this part of the thesis carves out the unique institutional characteristics of China's innovation system. This is followed by an account of *sociocultural* factors that influence the pursuit of innovation in China.

In order to address the research question that seek to increase understanding of innovation in Strategic Emerging Industries in China today, based on an evaluation from a historical perspective, the two main elements of research of this thesis are compared and integrated, resulting in new insights that extends our current perspective on innovation in China.

The illustration below outlines the initial research framework, which this thesis is based on. Guiding and structuring the research process, the framework is being qualified and extended based on the findings of the present study.

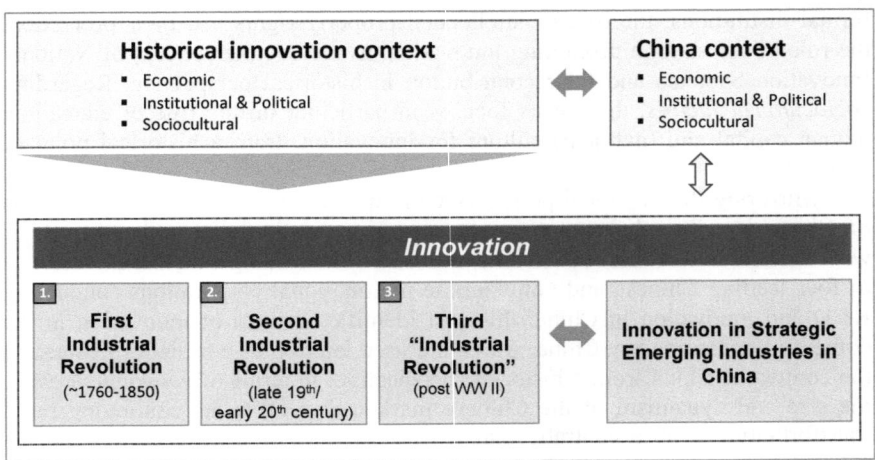

Figure 12: Initial research framework (own depiction)

4 Practical relevance of historical perspective

Overview: This chapter illustrates the practical relevance of this study. The first section 4.1 provides a managerial perspective on R&D and innovation in emerging markets and in particular in China, highlighting the challenges and shortcomings of Western MNCs when operating in China. Subsequently, section 4.2 explains why a more comprehensive historical perspective is relevant from a managerial perspective. Finally, section 4.3 provides a brief summary of the above points.

4.1 Innovation in emerging markets and China

In the face of new areas of innovation, e.g. in green technology, scholars often refer to unprecedented challenges that firms need to overcome in order to develop and market radical innovations successfully. Furthermore, as firms are increasingly locating their R&D centers in emerging markets, new challenges of cross-border R&D activities have emerged for multinational companies (von Zedtwitz, 2004). China in particular has become one of the most significant locations for R&D, featuring, the largest consumer market among all emerging economies, as well as a large and growing pool of university graduates as a potential talent pool for R&D activities. This has prompted an increasing number of multinational companies to move R&D and innovation related activities to China (Demirbag & Glaister, 2010; Sun et al., 2007).

However, despite the great potential of the Chinese market as a location for multinational companies' R&D units, companies are still facing significant challenges in conducting R&D in China.

A recent survey by the US-China Business Council in 2014 suggests that despite the opportunities that the Chinese market offers for foreign companies, there remain important challenges (USCBC, 2014). This survey confirms recent contributions on R&D management in China (e.g. Gassmann et al., 2012 and others, as outlined in chapter 3), which find that from a managerial perspective, these challenge concern in particular the *increasing competition* from Chinese companies in China; a lack of comprehension of China's *institutional and sociocultural context*; as well as *human resource* related issues.

First, the growth and development of China's national economy has been reflected in *increasing competition* from Chinese companies in China. While competition in itself was not perceived as negative in this survey, respondents did suspect that their Chinese competitors' increasing competitiveness may be based on favoritism by Chinese policymakers. Indeed, 67 percent of respondents "suspected, but [were] not certain" that Chinese state-owned competitors were receiving tangible benefits. Similarly, 29 percent stated that they knew that their Chinese state-owned competitors were receiving tangible benefits. Only 4 percent of respondents stated such kinds of support schemes did not exist (USCBC, 2014). Furthermore, previous studies have pointed out that in some industries, e.g. cement, steel and chemicals, *overcapacity*, due in part to fixed prices in state-owned enterprises, has impacted the market. However, as China is constantly moving towards market-based prices, e.g. for energy, this problem should gradually disappear.

Second, issues related to the institutional and sociocultural context of China are still a major impediment for foreign companies in China. Some of these issues can indeed be traced back to the innovation and industry policies implemented by the Chinese government in order to support the development of indigenous innovation, in particular in Strategic Emerging Industries (as outlined earlier). For example, the *enforcement of intellectual property rights* (IPR) in China remains a key issue. In the survey mentioned above, 91 percent of participants stated that they were "somewhat" or "very concerned" about IPR enforcement; of these, 48 percent were "very concerned" (USCBC, 2014). Other issues that foreign companies face in China include *foreign investment restrictions* and *licensing* issues. Importantly, while the Chinese institutional environment often represents a challenge to foreign companies in itself, foreign managers in China often feel that there is a *lack in transparency* and *insecurity* about local laws and regulations, and that the *enforcement and implementation* of the latter can be uneven, or *discriminating against foreign companies*.

Third, with regards to *human resources* related issues, R&D managers in China have pointed out *high employee turnover rates* and *low levels of individual initiative and innovative mindset* among their Chinese employees (Gassmann et al., 2012, p. 9). Furthermore, as wage levels in China have increased dramatically in recent years, from a managerial point of view, conducting R&D and innovation in China oftentimes does not result in significant *cost advantages* anymore. According to the study by the US-China Business Council, 90 percent of managers are concerned with rising human resource costs. However, only few companies seem to reduce or stop investment in China as a consequence, as these issues seem to be offset by the opportunities that the Chinese market offers (USCBC, 2014).

Therefore, due to the opportunities that the Chinese market offers for foreign companies to conduct R&D, as well as the difficulties that foreign companies have encountered in operating effectively in the Chinese context, a fine-grained and comprehensive understanding of the Chinese innovation context is essential for scholars and managers alike. To this end, a number of previous studies, outlined in chapter three, have served to increase understanding of China's business environment.

Building on these previous contributions, this thesis therefore seeks to make its main contribution to the literature on R&D and innovation management in China by including a historical perspective, which allows managers to assess the current situation and future developments in China in a more informed and realistic way.

The historical perspective is applied to the recently emerging phenomenon of Strategic Emerging Industries in China, to better understand how the economic, institutional and sociocultural characteristics of the Chinese context affect companies from a historical perspective. In particular, it includes the perspective of Chinese companies, which has been neglected in previous studies, due in part to limited access to Chinese company-level evidence. From a managerial point of view – both in China as well as abroad – knowledge about recent developments in China's innovation approach, as exemplified by its SEI, is highly important as a basis for current and future (e.g. R&D investment) decisions related to China. The following section outlines in greater detail how this historical perspective can add value to managers operating in China.

4.2   The relevance of history for management

Managers have recognized the growing importance of China as an innovation hub, as the evidence of growing investment in R&D activities in China clearly demonstrates. However, knowledge of China's increasing importance for innovation has not been accompanied by a better understanding of current developments in a larger context. As global business does not occur in a temporal vacuum, but instead can be seen as an ongoing thread of decisions made in the past, from a managerial point of view, the integration of history into management, and in particular with respect to innovation in China, is valuable for three main reasons.

First, understanding the past means to better understand the present and future. According to Knowles (2004), a historical perspective can provide a much needed context for managers in determining their organization's position by making comparisons with the past. Thus, managers can evaluate whether current events are part of a continuous trend or whether there are discontinuities. Similarly, it is an important tool for understanding human nature and its past endeavors and it can inform the present and future, while outlining economic development, as well as its industrial structures and the evolution of business strategies (Knowles, 2004). As outlined in chapter three, in reframing contemporary management issues with a view to its long-term meaning, the historical perspective offers a multi-disciplinary approach which can illuminate the relations between government and businesses, technological innovation, corporate culture and business ethics (Knowles, 2004), which can help managers operating in dynamic business environments in making more informed decisions based on a long-term perspective.

Second, integrating a historical perspective can help *identifying current approaches in management as an evolution from past approaches* that are subject to change in different time periods, showing their relative as compared to absolute nature. Every time period has unique sets of paradigms, which govern the discourse on issues of management and international business, such as globalization. Knowledge of history means an ability to operate beyond current meanings of management discourse. For example, approaches have included a discussion of management developments within a particular chronological period; the identification of various schools of thought in management and the demonstration of management theory and practice as a direct reflection of the ideas which emerged from these groups.

Third, an *awareness of historical relationships can be seen as a managerial skill*. The study of history in management can increase the ability to evaluate evidence and to develop an appropriate level of skepticism towards popular opinion and propaganda. It can also enable managers to better reflect on their decision-making process, by granting insights into human behavior operating under a different constraints and influences (Knowles, 2004). The historical perspective can also be seen as a way of thinking – oftentimes a search for patterns – and enable managers to ask the right questions – the "how", "why" and "what" questions that are typical of historical studies. For managers who are making decisions based on pre-defined strategies and principles, history offers "portrayals of reality against which those principles may be tested and experienced vicariously" ("Why History Matters to Managers," 1986). Lastly, an awareness of historical relationships can help managers make decisions in situations of ambiguity, by offering an alternative way of thinking that helps in accepting ambiguity as an omnipresent fact, to be comfortable with it, and to reject formulas ("Why History Matters to Managers," 1986).

For these three main reasons, for managers, a reflection about the past is necessary in order to better understand current events, as well as to make more qualified predictions about the future. Relating to an issue of such significant importance as innovation in China, this ability attains even greater relevance.

4.3 Summary

In summary, the present research topic is of significant practical relevance for two main reasons.

As the locus of global R&D has been shifting to emerging markets in recent years, and particularly to China, a more fine-grained understanding of the unique context of China is crucial in effectively managing innovation in China. Indeed, many of the most significant challenges that foreign companies face in China can only be tackled with a greater and deeper knowledge of the economic, institutional-political, as well as sociocultural context of China. From the perspective of (Western) foreign companies, these include issues related to competition landscape in China; insufficient legal transparency and enforcement, e.g. related to intellectual property rights; as well as challenges related to human resources in China. Therefore, a better understanding of new developments in China's innovation environment (e.g. related to indigenous innovation policies) is important in order to increase managers' ability to manage innovation in China more effectively.

However, as previous contributions have tended to neglect the historical dimension of innovation in China, this study seeks to provide a reference framework for managers in understanding and evaluating current developments as a consequence of past events and a precondition of the future. Reframing current approaches to management and identifying evolving schools of thought in management based on changing temporal paradigms, this can help managers deal with greater confidence with ambiguity and insecurity, which are frequently encountered in emerging markets such as China.

The following illustration summarizes these considerations.

PRACTICAL RELEVANCE OF RESEARCH	
1. Innovation management in China	**2.** History in management

• Rapid <u>increase in R&D</u> in China
• However, currently <u>a lack of understan-</u>
<u>ding</u> among foreign companies <u>about</u> how to operate in the context of <u>China</u>
• <u>Evolving China context</u> (e.g. SEI policies)
• Foreign firms in particular facing <u>unique</u> <u>challenges</u> in the context of China, e.g.
  • Increasing competition
  • IP protection issues
  • Investment related policies
  • Inssufficient legal enforcement and transparency
  • Human resources (e.g. educational system; rising costs)
	• <u>Understanding the past</u> to better understand the present and future
	  • Current trends as (dis-)continuities
	  • Reframing to long-term meaning
	  • Multi-disciplinary approach
	• Current <u>approaches in management</u> as an evolution from past approaches
	  • Identify contemporary paradigms
	  • Schools of thought in management
	• <u>Awareness</u> of historical relationships <u>as</u> <u>a managerial skill</u>
	  • Human behavior under different constraints and influences
	  • Dealing with ambiguity

Figure 13: Practical relevance of research (own depiction)

5 Empirical findings

Overview: This chapter presents the empirical findings of the present study. The first part – section 5.1 - presents a historical perspective on innovation, outlining the economic, institutional and political, as well as sociocultural factors that have been identified in the literature as conducive to or deterring innovation in the three distinct time periods outlined earlier. It also identifies historical factors responsible for shifts of innovation centers towards new regions and industries, providing insights into currently occurring trends.

The second part – section 5.2 - provides the empirical findings based on four case studies relating to innovation in China's Strategic Emerging Industries. For this, the initial reference framework introduced earlier provides the foundation for the presentation of the case studies and the empirical basis for the subsequent cross-case analysis and theory expansion. The new insights gained result in the theoretical and managerial implications and recommendations outlined in subsequent chapters.

5.1 The historical context of innovation

This section seeks to provide an understanding of how the context of innovation, marked by the economic, institutional, political and sociocultural environment factors, has evolved from a historical perspective, and how we can evaluate the context of innovation in China today based on the insights gained from this historical perspective, considering the three time periods provided earlier. The rationale for choosing these specific time periods and regions was provided in section 2.3 outlining the sample selection.

To recall, *economic factors* that were identified as affecting (national) levels of innovation include market size, as well as factor endowments, e.g. natural resources. *Institutional and political factors* include primarily formal and informal institutions, the role of intellectual property rights and their protection, the role of the state in promoting innovation, as well as the concept of National Innovation Systems. *Sociocultural factors* encompass primarily the level of education, human capital and (national) culture for innovation.

5.1.1   Innovation during the First Industrial Revolution

Before outlining the relevant factors that describe the innovation context in the period of the First Industrial Revolution, it is important to note that the term "Industrial Revolution" has been contested by a number of historians. While important innovations emerged during this time periods, some scholars rightly claim that the term "Revolution" may not adequately reflect the time period in all of its dimensions. For example, Nick Crafts has shown that economic growth was especially slow during the early period of the Industrial Revolution, with total factor productivity rising by less than 1 percent annually. Instead, he notes that the "hallmark of the Industrial Revolution was the emergence of a society that was capable of sustained technological progress and faster TFP growth. Future attempts by growth economists to model this transition should pay serious attention to the reasons for this improvement in the capability to generate innovations " (Nicholas Crafts, 2005, p. 533). As the time period that is referred to as the "Industrial Revolution" in the orthodox literature brought about an increasingly structured innovation system and can indeed be seen as an economic, political and social watershed, this thesis follows this terminus.

The period from around 1760 to roughly between 1820 and 1840 has come to be known as the First Industrial Revolution, or alternatively as the early period of the Industrial Revolution. Originating in Britain and spreading to North-Western Europe and later to the United States, this time period was marked by significant technological, organizational and institutional change across a number of sectors of the economy, fuelled in particular by new technologies in steam power, textile production, and iron making.

The first decades between 1770 and 1800 were characterized by a major technological transformation, with coal increasingly replacing charcoal and water power as a main source of energy. As coal needed to be transported, a network of canals was constructed around 1790, which in itself became a lucrative business. Coal-fired steam engines provided energy for rotative motion in manufacturing and transportation and was a key driver of technological transformation in this time period (Ayres, 1990, p. 3).

Much of the evidence about innovation in this time period is based on patent statistics. According to MacLeod, innovation in this time period increased significantly in the years after 1750. In particular, there was a strong rise in the number of patents for power source and textiles related equipment (MacLeod, 2002). However, patenting also grew for other capital goods including agricultural equipment, shipbuilding, canal building, and metallurgy (Fagerberg et al., 2006, p. 352). The following sections provide a more detailed qualitative perspective on the development of two key industries – textiles and steam power

– to illustrate the characteristics of the context of innovation during the First Industrial Revolution.

In *textiles*, there was significant increase in output, productivity and capital starting in the late 18th century. The introduction of mechanized production tools based on steam power considerably increased the productivity of factors laborers. For instance, the use of the power loom increased the output per worker by a factor of more than 40. Similarly, the cotton gin increased productivity of removing seed from cotton by a factor of 50. Large gains in productivity also occurred in spinning and weaving of wool and linen, but they were not as great as in cotton (Ayres, 1990, p. 17). Based on these new technologies, the price of cotton went down significantly and made it more available to the public.

Before the 1760s, textile production had mainly occurred in family structures, using flax and wool and targeted primarily at domestic consumption, as well as under the putting-out system, in which families completed products for central agents or subcontractors in their off-site private homes. Increasingly, the production of cotton textiles became popular, and cotton production in individual households rapidly increased, processing garments from overseas colonies such as India. Due to the increasing rivalry that resulted in relation with local manufacturers, the British state passed several bans and restrictions on the import of textiles into England to protect domestic industries. These came to be known as the Calico Acts (1690-1721) and were followed by the restriction of sale of most cotton textiles. However, due to the popularity of cotton, these restrictions only had a limited effect on stopping the rise of cotton as a dominant ingredient in textile production.

The following section describes how emerging centers of invention and innovation in the British textile industry emerged and which role individual business people and innovators played in driving the emergence of this industry. Increasingly, the production of cotton was concentrated in a small number of agglomerations such as Lancashire, strategically located near rivers for transport and featuring a vibrant network of production sites, workers as well as vendors.

In the 1760s, James Hargreaves invented the "Spinning Jenny" (a multi-spindle spinning frame) leading to significant improvements in thread production. Hargreaves had observed that the high demand for cotton necessitated a more productive spinning technique. At first, he kept the machine secret, producing only a small number of machines for use in his own cotton production. Later, he set up a shop producing spinning machines for other producers also. In July 1770, Hargreaves filed a patent (no. 962) on his invention, to preempt the use of counterfeits by other manufacturers. They eventually settled with a private agreement on payments (Espinasse, 1877). The case of James Hargreaves illustrates the dominant role of individual inventors in creating those innovations such have been seen as decisive drivers of development during the Industrial Revolution. Indeed, prior research has shown that suitable national institutional framework indeed are correlated with productive entrepreneurial activity (Baumol, 1996).

The example of Hargreaves, as well as other entrepreneurs of the First Industrial Revolution, suggests that it was mainly individuals' inventive and innovative efforts in developing new production devices and in integrating them in the shop-floor production process that resulted in the important innovations during this time period.

Another illustration is the case of Richard Arkwright (1732-1792), who invented the spinning frame, which was later developed into the water frame. The invention of the water frame resulted in two main advantages. First, it led to quality improvements in the yarn, so that linen and cotton were no longer the only materials that could be used for the production of fibers. Secondly, it resulted in spinning activities moving to larger scale facilities in areas with rivers that could produce energy to run more efficient equipment. Through this device, Arkwright was able to combine power, machinery, semi-skilled labor and cotton to create mass-produced yarn. Apart from his technical invention, Arkwright devised a number of organizational innovations, which greatly improved operational efficiency and made major contributions to the creation of the modern factory system (Griffiths, Hunt, & O'Brien, 1992).

As his firm was growing rapidly, employing 600 workers by 1774, Arkwright sought to secure the benefits of his invention and in 1775 filed for a great patent covering important processes of his invention, in order to attain monopoly power in a rapidly growing industry. However, as the public opinion in Lancashire was against the use of patents guaranteeing exclusive rights, he was not able to maintain his original patent from 1775, as courts claimed that his technology too closely resembled the spinning machines developed by Thomas Highs, a cotton manufacturer from Leigh, Lancaster (BBC News, 2015). This shows that the institutional environment in Britain at the time, already possessing a relatively well-established patent system, encouraged and honored individual entrepreneurs in developing new inventions and innovations. However, given the existing anti-monopoly policies, innovators could only reap the benefits of their work for a pre-defined time period and under certain conditions.

Most importantly, Richard Arkwright represents a class of entrepreneurs from the time period of the First Industrial Revolution in Britain, who were able to build the bridge from (technical) inventions to innovations, the latter of which could applied and diffused in the marketplace. In the case of Arkwright, other inventors (e.g. Thomas Highs) may have developed machines that formed the basis for Arkwright's commercial success. However, it was Arkwright who nurtured and financed these inventors, patented ideas, and protected the machines. He can therefore be seen as an early day example of a private investor in innovation.

The case of *steam power* provides another case in point that illustrates the economic, political and sociocultural innovation context of the First Industrial Revolution. The emergence of the stationary steam engine can be seen as another critical innovation of this time period. However, it is important to note that the steam power engine diffused relatively slowly and had only limited effects on economic growth, as well as in terms of linkages to other industries (von Tunzelmann, 1978). In 1775, James Watt took his steam engine to the market. Nevertheless, it diffused rather slowly: in 1800, twenty-five years after its introduction, there were only 21 engines in Manchester, which had emerged as a major center of textiles by that time. According to von Tunzelmann (1978), technology emerges slowly when costs are high: for many decades after its introduction, this was true for steam engines, as well as the necessary fuels to operate them.

The following "Historical case in point" provides an illustrative example of the development of the steam engine by James Watt (1736-1819), a Scottish inventor and mechanical engineer whose improvements to previous types of steam engines – most notably to the Newcomen steam engine – had a substantial impact on the economic and social transformation of Britain and subsequently in other countries.

Table 4: Historical case in point A: James Watt in the 18th century

Historical case in point A: *James Watt's steam engine in the 18th century*

Despite common perception, James Watt, a Scottish inventor and mechanical engineer who lived in Birmingham in the 18th and 19th century, was not first to invent the steam engine. Thomas Newcomen, a Baptist preacher from Devonshire, England, had previously built the first practical steam engine for pumping water from mines in 1712, which remained in use for almost fifty years for that purpose. However, James Watts's achievement was to further refine the energy-inefficient previous steam engines by using a separate condenser, greatly improving the power, efficiency, and cost-effectiveness of steam engines. This turned the steam engine into a disruptive technology that was fundamental to the Industrial Revolution, as it enabled production in the newly emerging factories.

Nevertheless, the technological inventions developed by James Watt would most likely not have been turned into groundbreaking, large-scale innovations without the influence of several influencing factors. Most importantly, his business partner and financier, Matthew Boulton, served as an entrepreneur, turning Watt's invention into a commercially attractive innovation, as the success of the Boulton-Watt engine shows, which served as a major power source of the Industrial Revolution. The roles of inventor and entrepreneur that characterized the Watt-Boulton relationship is apparent from their written exchanges, in which Watt laments about his "inability to bargain and struggle for my own with mankind; all which disqualify me for any great undertaking" (Watt's letter to Roebuck on September 24, 1769), while businessman Boulton was "excited by...love of a money-getting, ingenious project...to...produce the most profit...It would not be worth my while to make for three counties only; but I find it very well worth my while to make for all the world" (Boulton's letter to Watt on February 7, 1769).

At the time, the institutional environment of Britain supported this productive type of entrepreneurship enabling the steam engine venture. Letter exchanges between Watt and Boulton suggest that there was a high degree of insecurity related to the development of their venture, with significant amounts of upfront investments and sunk costs required. The existence of reliable patent and litigation systems in contemporary Britain and thus the ability to collect royalties on the use of engines (which initially represented most of the firm's profits) allowed the Boulton-Watt company to pursue an expansionary investment policy. In addition, Cain and Hopkins have pointed out to the importance of "gentlemanly ideals...[which] provided a shared code, based on honor and obligation, which acted as a blueprint for conduct in occupations whose primary function was to manage men rather than machines" (Cain & Hopkins, 1993, p. 26).

The Boulton-Watt venture can therefore be seen as an early example of transformative innovation, which was enabled by a combination of supply-side factors – most significantly the entrepreneurial role of Boulton – and several demand-side factors. In particular, the quality of legal and financial institutions encountered in Britain at the time should be pointed out. The latter secured physical and intellectual property rights and gave entrepreneurs such as Boulton and Watt the security and incentives to make necessary investments. Although the development of the steam engine inferred significant costs, it was funded through a personal network of individual investors such as Boulton, rather than through loans given out by banks or formalized venture capital funds.

The example of the Boulton-Watt steam engine shows that Britain in the late 18th century was first among European countries to feature highly developed institutions allowing for individuals to engage in "productive entrepreneurship" (Baumol, 1996) leading to innovation. In contrast, unproductive entrepreneurship would result in rent seeking, organized crime or other redistributive activities. Once these institutions were in place, entrepreneurs in different industries were more likely to feel impelled to pursue opportunities for commercial purposes, creating an environment that propelled further innovations and fueling industrialization in Britain. The example of Boulton-Watt also shows the importance of kinship and relationships that were particular important at a time when institutionalized innovation systems had not yet emerged.

Indeed, as early as during the first century AD, Heron of Alexandria had already developed initial versions of the steam engine. However, it was merely used for "amusement" (J Mokyr, 2012, p. 532) due to limited opportunities for non-military, industrial or commercial application and a lack of entrepreneurs to commercialize his invention. This suggests that significant innovation emerged in parallel with supportive institutional structures. In this context, although individual entrepreneurs have existed at all times and across regions, previous studies have shown a significant correlation between institutional quality and productive entrepreneurship (e.g. Baumol, 1996; Sobel, 2008).

Thus, the system of innovation of the First Industrial Revolution is characterized by a focus on crafts and production related innovation, as well as improvements in the efficiency and organization of the production process. This "Industrial Enlightenment" resulted in better knowledge of industrial and artisanal techniques, which were recorded in manuals and handbooks on industrial practices (Joel Mokyr, 2002, pp. 34–35).

In describing the sociocultural context of this time period, Mokyr has argued that in 18th century Britain, a "new enlightenment ideology" had emerged, in which the agricultural and commercial elites were convinced that "the economic game was not zero-sum and that a free-market environment of open access, competition, and unrestrained innovation was the patriotic and virtuous thing to do. As it turns out, it was also the profitable thing to do"(Mokyr in D. S. Landes, Mokyr, & Baumol, 2012, p. 202).

Therefore, both demand and supply factors can explain the increase in innovative activity in this time period. In terms of demand factors, the willingness of individual entrepreneurs to take financial and operational risks in venturing into new businesses – in the absence of public sources of financing – should be highlighted. In terms of supply factors, institutional changes, the improvement of intellectual property rights and legal and financial stability provided a suitable ecosystem for innovation and a basis for increasing prosperity and the emergence of consumer good markets. Therefore, both the ingenuity and ambition of individuals, as well organizational improvements were crucial in allowing for innovation during the First Industrial Revolution.

The scale and organizational complexity grew further in subsequent decades. Increasingly, new technological innovations required investments on a larger scale (e.g. such steamboats and railroads), making formal financial institutions increasingly important and leading to the emergence of increasingly large and sophisticated financial institutions and often a separation of inventor, entrepreneur and financier. Due to the increasing scale of new enterprises, the management of organizations became increasingly complex.

This development is illustrated well by the railroad industry, which emerged first in Britain and then in the United States in the 18th and 19th century. Originating in earlier innovations in the area of steam engines and metallurgy, the railroad industry provides a suitable gateway connecting the First Industrial Revolution – as outlined above, with the "Second Industrial Revolution", which is outlined in the subsequent section, describing in particular the developments in the electrical equipment and chemicals industries in Germany and the United States.

Table 5: Historical case in point B: Railroads in the 18th and 19th century

Historical case in point B: *Railroads in the 18th & 19th century*

The emergence of railroads in Britain and the United States in the 18th and 19th century provides another illustrative example of innovation, which increasingly replaced other forms of transportation such as carriages and canal ships only after its commercialization and more wide-spread diffusion from the early 19th century.

In 1804, Richard Trevithick, an English engineer from Cornwall, built the first full scale working railway steam locomotive in the United Kingdom and can be credited for having invented the locomotive. However, it was George Stephenson and his son Robert who turned the invention into a true innovation, founding the Stockton and Darlington Railway Company in 1821 and later the Liverpool and Manchester Railway ("L&MR") from 1830, the earliest example of a public locomotive train connecting urban areas. He also established the "Stephenson gauge" which has been the world's standard gauge to this day (Savage & Barker, 2012).

Initially, railways were used to connect coalmines to the nearest port, river, or canal dock, where the cargo would be transferred to water (J Mokyr, 2012, p. 233), resulting in a patchwork of numerous and unconnected railway links. In 1830, the world's first high-speed railway was opened, connecting urban centers and carrying both freight and passengers. The success of this line led to increasing investments into railways. While earlier railway projects were mainly financed by a small number of banks and wealthy individuals, newly emerging railway companies such as the Liverpool and Manchester (the "L&M") increasingly attracted investments by members of the newly affluent middle class created by the Industrial Revolution. Motivated by commercial interests, from the 1830s, a large number of railway companies were set up through Acts of Parliament. Increasing railway share prices resulted in increasing investments, until this "railway mania" peaked and collapsed in 1846. In this context, the government took a laissez faire approach to the construction and operation of railways, leaving the development of Britain's (and the world's) first high-speed railway network from the 1830s largely in the hands of individual entrepreneurs and especially consultant engineers who planned increasingly well-connected links.

In the United States, railway construction started in the 1830, closely following and copying British railroad technology, with British railroad engineers often leaving Britain to take up lucrative opportunities in the US railway industry, leading to a significant transfer of knowledge. From the 1870s onwards, the role of railway entrepreneurs and financiers increasingly shifted from merely building new railways to consolidating and restructuring existing networks. For example, in 1885, New York based financier J.P. Morgan leased several railroads (including the New York and Buffalo Railroad) to the New York Central (Carosso, 1987, p. 331). Furthermore, from the 1850s onwards, the federal government became involved in supporting the development of railroads, by giving out land grants and subsidies to railway companies to open railway links to the Western parts of the country (e.g. the Union Pacific-Central Pacific), as well as by aiding war efforts, with the Confederate government taking railroads and their rolling stock under military control in mid-1863 (Massey, 1952). Despite the importance of the railway expansion in the United States in the 19th century, there has been severe criticism about the unethical treatment of Native Americans, many of whom were mistreated, displaced from their lands and confined to designated reservations, in order to allow for the westwards expansion of the US railway network. This aspect is further developed in section 6.1.1.

As Chandler has demonstrated, the size and dynamism of the US market provided an important impetus for large-scale investment in the newly developing railroad industry in the 19th century, as well as providing the foundation for the emergence of large integrated firms (Chandler, 1990).

In comparing the innovation context in which steam engines and railroads emerged, the increasing maturity of institutions as demand-side factors is most striking. Compared to the development of the steam engines around sixty years earlier, at the time of the first railway networks in Britain, institutions had become more mature and a significant middle class had emerged, providing a source of entrepreneurs, as well as investors. In the United States, political motivation to drive westwards expansion was a major driver of the rapid expansion of railroads – often times at the expense of American Indian tribes who were displaced and confined, sacrificed for the greater goal of expanding national railway networks.

For the development of the steam engine as an industry-disrupting technological innovation, the function of elementary institutions – namely the protection of physical and intellectual property - provided an environment in which some inventors had the opportunity to develop commercially attractive innovations with the support of wealthy individual investors. Increasingly, formal financial institutions, such as private banks and government institutions, replaced informal financial networks. Interestingly, the importance of informal networks for access to financial capital remained important into the 20th and 21st centuries in regions where formal institutions are lacking, as recent studies show, e.g.

Wolcott (2010) in the context of post-1948 India, and Rauch and Trindade (2002) in the case of ethnic Chinese networks in international trade.

The evolution of institutional demand-side factors of innovation, using the examples of Britain and the United States in the 18th and 19th century, seems to suggest that the mere existence of functioning legal and economic institutions provides incentives to inventors and entrepreneurs to engage in productive entrepreneurship, even in the absence of formal financial institutions. Therefore, in accounting for the evolution of factors enabling such disruptive innovation, one can observe a gradual shift in the importance of demand and supply side factors. The evidence suggests that the role of the individual entrepreneur – e.g. social and educational, ethnic and religious background, character and values - remains central to the emergence of disruptive innovation.

One can observe that over time, increasing levels of institutional sophistication increase the relevance of demand-side factors of innovation, e.g. the quality of legal and political institutions and the protection of intellectual property rights and patents, as entrepreneurs increasingly rely on them and adjust their entrepreneurial strategies accordingly. In contrast, in the US and Western Europe before the start of the Industrial Revolution, as well as more in regions with less developed institutional frameworks even up to this day, we can observe a relatively more important role of the individual entrepreneur who may innovate despite rather than supported by the operating environment, forcing him or her to make up for missing prerequisites, in a similar vein as envisioned by Gerschenkron's notion of governments who can substitute for "missing prerequisites" to trigger economic growth, e.g. by developing financial mechanisms to support entrepreneurship and thus escape "economic backwardness" (Gerschenkron, 1962).

Therefore, one may argue that demand-side factors, in particular institutions, have become more relevant for the development of disruptive innovation not only because entrepreneurs have a greater need for them, but also because they have become more available, making them part of innovator's calculation and thus increasingly indispensable for innovation. This suggests that the evolution of factors that make up the innovation environment is neither linear nor one-directional. Instead, its different elements are dynamic and mutually dependent. Their relative importance can be imagined as a pendulum, swinging back and forth based on the requirements of the envisioned innovation project, as well as the currently existing set of external factors, which mutually influence each other over the long-term.

In conclusion, the environment that allowed innovation to first occur on a broad scale in Britain (and later on in the United States) was based on an institutional framework that provided suitable incentives for innovators, and was complemented by a relatively high level of human and natural (most importantly coal) factor endowment, and embedded by government policies that supported this positive cycle of increasing innovation, productivity and economic growth. The innovation system that was characteristic of the First Industrial Revolution was based on the entrepreneurial initiative of individuals, an orientation towards crafts such as textiles, and little automation, with knowledge in wood- and metalworking was valuable. While innovation in this time period was driven to a great degree by rising demand (e.g. for textiles) by an emerging market of consumers, the institutional framework of – as exemplified by Britain – provided relative security and stability as a basis for entrepreneurship, supporting further economic growth.

5.1.2 Innovation during the Second Industrial Revolution

This section describes the system of innovation that emerged during the "Second Industrial Revolution". As noted earlier, the term "Second Industrial Revolution" used in this study refers to the time period between the late nineteenth and World War Two, describing a second wave of technological and organizational progress and differing from the common definition of the Second Industrial Revolution as the second phase of the Industrial Revolution. In the early years, this time period was particularly characterized by the continuing expansion of railroads, large scale iron and steel production, the diffusion of new types of machinery in manufacturing, increased use of steam power, use of oil, beginning of electricity and by electrical communications. Therefore, this time period has often been referred to as Technological Revolution.

In particular, the developments in the electrical equipment and chemicals industries in Germany and the United States illustrate the increasing role of industrial R&D in large firms. In the United States, its large and increasingly integrated market provided a strong stimulus for this type of innovation.

In the time period starting in the late nineteenth century, Britain's previously most dominant industries such as iron, coal, textiles, and steam engine technologies, became relatively less important. Instead, new technologies and industries emerged including petroleum, and later on chemicals, optics, and eventually electricity. Furthermore, in this period, the geographical center of innovation and technological leadership gradually shifted from Britain to central Europe – Germany and France in particular – as well as the United States (Fagerberg et al., 2006).

During this time period, the size of firms grew, leading to changes in the firm structure and related organizational innovations that continued to mark a newly emerging relationship between firms and research institutions, which continued to intensify in the twentieth century. For example, in the global dye industry in the late nineteenth century, German chemical dye companies became increasingly dominant, based on advantages in economies of scale and scope, representing 50 percent of global dye production by around 1870, and 75 percent of global production by 1900 (Murmann & Homburg, 2001).

These developments had important effects on the context of innovation. While innovation in the First Industrial Revolution had largely resulted from the ingenuity of individual entrepreneurs – embodied by entrepreneurs such as Hargreaves, Watt or Stephenson – increasingly, innovation would result from firm-internal or external resources, with formal education playing a greater role than before. Furthermore, scientific and technological knowledge was increasingly developed within firms, providing a major impetus for innovation. However, the shift from innovation from basic industries such as steam power towards modern industries such as chemicals was gradual. The following historical case in point about the Henry Ford Company shows that in the late nineteenth century, innovation increasingly focused on rapidly growing consumer markets; at the time, business expansion into the United States provided almost limitless opportunities for entrepreneurs. The economic structure of these newly emerging economic centers, featuring large and growing markets, can help explain the shift of innovation activities to these markets, as well as the resulting changes in the innovation process focused on standardization and consumer-centric, pragmatic consumer products.

Table 6: Historical case in point C: The Henry Ford Company

Historical case in point C: *The Henry Ford Company*

The history of the automobile - first as a product of technological novelty and later as a means of transport for the masses – provides an illustrative case on the importance of the personality of the innovating entrepreneur. The modern automobile goes back to a series of technological inventions: the first notion of an "auto-mobile" goes back to a Jesuit missionary who constructed the first vehicle, powered by steam, as a present for the Chinese Emperor (Golvers & Verbiest, 2003) around 1672; however, it could not carry a driver. François Isaac de Rivaz developed the first car with an internal combustion engine in 1807. In 1886, the German engineer Karl Benz created the first "Patent-Motorwagen", which can be seen as the beginning the modern automobile (Eckermann, 2001).

While the above inventions were radical innovations in terms of technological ingenuity, they were no innovations at a transformative scale. It was Henry Ford who created disruptive innovation, by converting the automobile from an expensive item of conspicuous consumption into a practical conveyance that was affordable to a large number of middle class Americans, with significant influence on the United States' landscape of the twentieth century. He achieved this purpose by relentlessly making the production of a single model – the Model T – more efficient, simple and inexpensive, lowering its retail price to US$ 825 in 1908, so that by 1920, a majority of Americans had learned how to drive.

Ford's ability to generate innovation can be largely attributed to his own ingenious capabilities as an engineer and entrepreneur, with a vision to make the automobile accessible for the masses and particularly to farmers, which were especially dear to him due to his rural upbringing in a Christian family. Therefore, his entrepreneurial activities were largely driven by supply-side factors of innovation: Henry Ford was trained as an engineer, which helped him understand the production process of his vehicles, resulting in Ford's factories becoming the world's most efficient at the time. Furthermore, his culture and religion can be seen as further influencing factors: brought up in a rural setting, Ford was a Freemason who was said to believe in reincarnation (Marquis, 1923). Although at the time of the development of his vehicles, there were already well-established financial institutions in the United States, Henry Ford always preferred to rely on financing from personal networks, such as John and Horace E. Dodge (known as the Dodge Brothers) as well as other friends and colleagues. Not until the 1930s did Henry Ford overcome his resistance to finance companies and reliance on external creditors. Important demand-side factors for Ford included the enormous potential size of the US market (whose citizen at the time predominantly lived in rural areas), with significant expected profits from innovation.

The case of Henry illustrates that despite evolving elements shaping the innovation environment – e.g. more mature political and legal institutions developing in one region over time – the emergence of innovation still needs to be seen in combination with supply-side factors, such as determined by the individual entrepreneur. Innovation does not necessarily only when the context is suitable, e.g. from an institutional point of view. Instead, individual ingenuity leading to significant innovation may occur not only because of a suitable environment, but also despite particular environmental factors. Therefore, while the environment of innovation clearly matters, individual entrepreneurship seems to matter just as much.

Indeed, in the automotive industry, Henry Ford was a pioneer in serving mass-market customers. The expansion of the car industry continued well into the middle of the twentieth century in Western Europe and the United States, and in some world regions it is still rising today. The growth of industrial firms resulted in larger, vertically integrated firms, which increasingly became the norm in Germany and the United States. In parallel, these firms integrated incorporated research and development departments or laboratories, in which teams of researchers – often in a cooperation network with universities – would conduct scientific work (Bruland & Mowery, 2004, p. 7). During the years up to the Second World War, firms started to formulate industrial innovation as an important component of corporate strategy. In the eyes of Schumpeter, professionally managed firms would become important agents of "creative destruction" (Joseph A Schumpeter, 1942).

An important driver of innovation and the development of new products during the late 19th and the early 20th century were increasingly large and integrated markets. In particular, the market size of the United States was increasingly conducive to commercializing innovation. However, besides physical market size, improvements in infrastructure and communications allowed for integration and the emergence of "national" markets, spurred also by the invention and diffusion of telegraphs and telephones in the period after the Civil War. Expanding railroads links reduced the cost of transportation of materials and products, which in turn reduced the price of materials to build further infrastructure. In the United States, falling transportation and communication costs greatly facilitated westward expansion and economic development, by connecting the western frontier with the political, industrial and financial centers of the East coast.

In the following, we focus on the *chemical industry* in Britain, Germany and the United States, as this industry epitomizes the major evolution in the context of innovation during the Second Industrial Revolution, providing important insights into the role of firm size and organization, the evolution of internal and external sources of R&D, and the role of the state in providing a suitable context for innovation.

The chemical industry emerging in the late 19th century had its origins in the dyestuff industry in the textile industries of Britain. Initially, the demand for larger quantities, more variety and better quality of dyes was met by searching for more efficient extraction methods of natural dyes and better techniques of attaching them to fabrics. In 1856, the English chemist William Henry Perkins discovered the first synthetic dyestuff, produced from the abundantly available coal tar, a by-product from the process of making coal gas and coke. The emergence of synthetic dyes in the middle of the nineteenth century can be seen as the last in a series of major innovations in the textile and related industries, giving rise to subsequent industries.

In the years starting in the late 1870s, demand for synthetic dyestuffs continued to rise and increasing numbers of patents were registered. In this time, the pattern of innovation shifted, with innovation resulting less from individual ingenuity and invention, but increasing emerging from institutionalized research and development. The subsequent time period was marked by the relative decline of the British dye industry and the rapid growth and increasing global dominance of the German chemical industry in terms of output as well as innovation. The success of Germany's chemicals industry has been highlighted by previous scholars as "Imperial Germany's greatest industrial achievement."

In debating the importance of demand as compared to supply side factors to innovation, most scholars now agree that the rise of the German chemicals industry was not due to greater demand in Germany for dyes, but instead due to supply side factors, with rates of scientific discovery rising rapidly in German after 1871 (Walsh, 1984). Much earlier than in other countries, a network of industrial R&D laboratories emerged in Germany, which operated in collaboration with publicly funded research and technical universities, providing education to young researchers and scientists at an increasingly large scale. These significant efforts resulted in a rapid rate of scientific discovery and innovation and to Germany's domination of the world chemical market.

The rapid expansion of German research universities, which took a pivotal role in providing training for industry scientists and engineers, and the focus on science and technology, can be seen as a result of deliberate government policies in Germany after its unification as German Empire in 1871 in order to stimulate industrialization and to catch up with early industrializing countries such as Britain.

According to Murmann (2000), by the 1870s, Germany had almost thirty university and technical university departments in organic chemistry, as well as seven large centers for organic chemistry research and teaching (Haber, 1971). British universities, in contrast, received much less public funding and were less closely linked with industry. While only 1,000 students at leading British universities were enrolled in engineering subjects in 1911, German technical universities enrolled 11,000 students in engineering and science degrees by 1911 (Haber, 1971). Similarly, while state support for university-level education was only about GBP 26,000 in Britain in 1899, in Prussia, which was only one part of Germany, the state provided GBP 476,000 in the same year – more than eighteen times as much – for this purpose. By 1911, these figures had evolved to GBP 123,000 and GBP 700,000 respectively (Haber, 1971). As a result, between 1886 and 1900, the six largest German chemical firms took out 948 dye patents, while the six largest British firms took out 86 patents in the same time period (Walsh, 1984).

Many policies to accelerate development were based on the ideas of Friedrich List, a leading German-American economist in the nineteenth century who advocated for the protection of infant industries and other policies to speed up industrialization. List's envisioned strategies were mostly related to knowledge acquisition to enable the generation, as well as application of new technology, recognizing the relationship between the integration of foreign technology and industrial development at home. Recognizing the importance of key technologies developed by British engineers in the nineteenth century, the government of Prussia was involved in acquiring British technology, regardless of significant resistance from the British government, who tried to prevent this technology export and imposed fines for its contravention (Freeman, 1995b). List therefore developed an early notion of the modern national system of innovation, including important features such as education and training institutions, the role of science, technical institutes, knowledge accumulation, the adaptation of imported technology, and the promotion of strategic industries. In particular, he emphasized the role of governments in devising and implementing industrial and innovation-related policies supporting national development.

In relating history to the present, one can clearly identify some parallels between the ideas of Friedrich List outlined above, and the economic policies that were pursued in China in the post-Mao period, which were aimed towards the development of the national economy and have included the promotion of strategic industries, as well as national education and training in a similar way as envisioned by Friedrich List. In a similar vein, the industrial and innovation policies pursued by China today have resulted in allegations of unfair treatment, in particular by foreign governments and businesses (e.g. USCBC, 2013).

In Britain, industrial leadership positions continued to be recruited among non-technical professionals. In contrast, in Germany, technically trained managers increasingly moved into manager positions, resulting in stronger ties between corporate strategy and industrial research (Bruland & Mowery, 2004, p. 8). The focus on technology education and development, supported by the German government, had a positive impact not only on the chemicals industry, but also on other industries such as the electrical equipment and engineering industry, helping the emergence of companies such as Siemens.

As government had provided a supportive context for the rise of these industries, their increasing scale and profitability also resulted in increased lobbying of the government to receive more research funding. In 1874, Werner von Siemens initiated the German Association for Patent Protection, resulting in the first national patent law in the newly founded German state being introduced in 1877 (Bruland & Mowery, 2004, p. 9). In 1887, von Siemens bestowed a large plot for the Imperial Institute of Physics and Technology in Berlin, in close proximity to Siemens' headquarter area. Subsequently, this research institution was built up with public funds in that year (Bruland & Mowery, 2004, p. 9). Stronger intellectual property protection in subsequent years helped firms in appropriating the returns to their R&D. After the passage of the new German patent law in 1877, many of larges German chemicals firms established formal in-house R&D (Bruland & Mowery, 2004, p. 9).

Therefore, institutional change can also happen when stakeholders in the innovation system – for instance, entrepreneurs and firms – actively seek to have influence. This illustrates the coevolving nature of innovation, in which the emergence of innovation depends on institutional arrangements, but also shapes the latter.

Table 7: Historical case in point D: Industrial research & innovation at Bayer

Historical case in point D: *Industrial research and innovation at Bayer*

In the years before 1870, there was only little research conducted in the dyestuff industry, with individual entrepreneurs (oftentimes chemists who started their own businesses or worked for existing firms) integrating new scientific knowledge and production methods into practice. However, science and innovation were not yet institutionalized. This changed in 1870s

Bayer can serve as a good illustration of this development. In the 1860s, Bayer's laboratories were still small and poorly equipped. In 1877, Bayer employed seven chemists, four of whom were occupied with routine analysis and two with color testing (Meyer-Thurow, 1982, p. 365). In subsequent years, progress was achieved mainly through trial-and-error chance discoveries. In addition, there was a high degree of mobility among Bayer's chemists, who were inclined to take up other private sector opportunities in teaching, consulting or entrepreneurship.

However, as the interest in dyestuffs and chemicals increased, new patent laws resulted in research moving increasingly from academic institutions towards industrial corporations, with other competitors such as Hoechst and BASF emerging (Meyer-Thurow, 1982, p. 368). Bayer intensified its relationships with local technical universities and founded an in-house research laboratory. While this laboratory, led by Carl Duisberg, was initially still involved with issues in production engineering and problem-solving, in 1891, Bayer finally established a dedicated laboratory, showing a clear differentiation within the firm between R&D and operational support (Bruland & Mowery, 2004, p. 9).

As a result of its increasing focus on industrial research, the number of German patents filed by Bayer rose from 36 in the period from 1877 to 1886, to 99 in the period from 1887 to 1890, to 512 in the period from 1891to 1900. The growth in patents shows the increasing level of research intensity that occurred during the late 1880s, as well as the stabilization of the research effort due to the foundation of the main scientific laboratory in 1891. By 1900, Bayer held more than 2500 German and foreign patents, and more than 8000 by 1914 (Meyer-Thurow, 1982, p. 381).

The establishment of significant industrial research also served as a barrier of entry to new firms seeking to enter the market. Indeed, between the mid-1880s and 1914, no new dyestuffs firm was successfully founded, with previous market leaders still dominant in 1914 (Meyer-Thurow, 1982, p. 381). A virtuous cycle of industrial research and innovation based on "expansion, integration, and diversification" strategies (Meyer-Thurow, 1982, p. 381) contributed to the development of firms like Bayer, which developed from small dyestuff companies into large chemical and pharmaceutical corporations, leading to the accumulation of a knowledge stock that has helped the company stay at the forefront of innovation to this day.

Therefore, the establishment of in-house industrial R&D departments in Germany from 1870 can be seen as a major institutional innovation with profound impact on how innovation emerged. While firms had engaged in product and process innovation already during the First Industrial Revolution, as illustrated in the previous chapter, the German dyestuff and later chemical industry led the way in establishing R&D for new products and processes on a larger and more systematic scale (Beer, 1959). The success of the German chemical industry resulted in firms in other European countries imitating the emerging social innovation represented by the R&D Department. Specialized laboratories for research and development therefore gradually became the industry standard in many manufacturing firms between the late nineteenth and the first half of the twentieth century. In the electrical industry in Germany and the US, this type of R&D emerged in the late 19th century, in companies such as Siemens and AEG in Germany, or General Electric and Alcoa in the US (Mowery, 1981).

In the United States, the use of external, contract-based research firms was more common than in Germany. For example, the first industrial R&D center of Du Pont, a US chemicals firm, was opened in 1902. It was the first laboratory that was separated from the manufacturing operations in terms of location and organization. However, throughout the twentieth century, the role of contract-based industrial research declined, due to the complexity and uncertainty of large research projects (Bruland & Mowery, 2004, p. 10).

In comparing the evolution of industry-based innovation in the United States to that in Germany, one has to differentiate between the quality of scientific research and the level of technological innovation and change.

First, in terms of scientific research in chemistry, Germany continued to be a world leader until the Second World War, ahead of the United States. For instance, through 1939, German scientists received fifteen out of the thirty Nobel Prize awarded in chemistry, while US scientists received three, and French and British scientists each accounted for six (Bruland & Mowery, 2004, p. 11). From the 1930s and especially after 1945, the United States gradually assumed a scientific leadership position. Between 1940 and 1994, US scientists received thirty six out of sixty five chemistry awards, while German scientists received eleven, British scientists received seventeen, and French scientists received only one (Bruland & Mowery, 2004, p. 11).

Second, in terms of technological innovation and change, from the beginning, the American chemical industry was based on specific features of the US market. On the one hand, the large size and dynamic growth of its domestic market made large-scale, continuous process production profitable. On the other hand, the existence of significant natural resources – such as oil and gas that were suitable ingredients in the chemicals industry – allowed US companies in the organic chemical industry to transform their resource base and resulted in significant cost advantages. In particular, the endowment in oil reserves - spurred also by increased demand for liquid fuels from the rapidly growing automotive industry in the early twentieth century – led to the development of a large petroleum refining industry. The large volume of production led to experience in capital-intensive manufacturing and continuous-process technologies, e.g. in firms such as Union Carbide and Standard Oil (New Jersey). In combination with low-cost petroleum and natural gas from US petroleum firms new and large process innovations developed by US firms resulted in significant manufacturing cost reductions (Bruland & Mowery, 2004, p. 11).

In Germany, firms developed new technologies to compensate for the lack of domestic feedstock, using synthetic gasoline as fuels and tires made from coal-based synthetic rubber. Only after the Second World War, with the revival of international trade and investment, some of these restrictions were relaxed and innovation and industry patterns in Europe gradually came to resemble those of the United States.

In conclusion, the evolution of the innovation system occurring from the late 19th century – most notably the increased importance of state-run supported R&D and contract-based research – suggests that the establishment of a formula for the systematic development of invention as a routine was the greatest novelty in this period.

The role of science and larger scale, industrial R&D remained central from this period onwards. Innovations like the radar, computers or rockets were products of this type of industrial research, integrating a network of stakeholders including governments, as well as industrial and academic engineers and scientists (Freeman, 1995b). In summarizing the transition from inventor-entrepreneurship towards in-house industrial research, Schumpeter argued that industrial research would strengthen, rather than weaken, the position of industry leaders (Schumpeter, 2013), which has been supported by data on research employment and firm turnover more recently (Mowery, 1983). In the US in particular, a large and homogenous market supported the emergence of institutionalized innovation in large companies.

5.1.3   Innovation since the Second World War

In the time period after the Second World War, innovation leadership moved from Europe to the United States. Furthermore, a number of economies outside of Europe and the United States experienced rapid (catch-up) growth, based on targeted innovation and development policies. In particular, the experience of post-war Japan, followed by the four "Asian Tigers" – Hong Kong, Singapore,

South Korea, and Taiwan – provides interesting insights on the evolution of the innovation system after World War Two and until today. In the following part, the evolution of the innovation system in this time period is illustrated, focusing on the experience of the United States in particular, as well as Asian economies.

The evolution of the innovation system after the Second World War needs to be seen in the context of the experience of the War, as well as the beginning of the Cold War. During the war, governments had become important actors in the system of innovation, by driving demand for military-related industries. A good illustration of this is the Manhattan Project, a publically funded research project that resulted in the development of the first atomic bombs between 1942 and 1945. The nuclear devices developed in this project were later used in the atomic bombings of Hiroshima and Nagasaki, Japan in 1945. With the outbreak of the Cold War from about 1947, public investment in R&D sharply increased, motivated largely by national security reasons.

The evolution of the innovation system in the *United States* after the war was marked in particular by three distinctive features: the large share of government in national R&D, dominated by expenditures related to defense; the important role of antitrust policy; as well as the important role of small, new firms in developing and commercializing new technologies, often in collaboration with larger firms.

The share of government spending in R&D increased dramatically in the United States of the postwar years. Before the war, corporations had become major sources of research and development and innovation, in terms of funding as well as development. This changed significantly following the war. Public funding of R&D increased dramatically, with the share of the federal government in national R&D spending rising from 12-20 percent in the 1930s to 40-50 percent of national R&D in the postwar period (Mowery, 1994, p. 88). However, in contrast to other industrial economies, a relatively large fraction of federally financed research was performed in non-government research laboratories. Military services continued to dominate the national US budget for R&D starting in the early 1950s, decreasing to less than half of the budget in only three years in the time period between 1960 and 1990 (Mowery, 1994). However, postwar US military procurement was more important for the emergence of high-tech companies as compared to R&D expenditures related to defense. This is because profits from military procurement contracts provided funds to company-funded R&D and generated more civilian spillovers than R&D that was directly funded by the military. ; one example of this is the early American semiconductor industry.

Similarly, US government procurement remained an important pillar of support for newly emerging industries. For example, IBM's projected sale of fifty (out of a total of 250 projected global sales) computers had an impact on the company's decision to develop its first device for businesses, labeled as "650" (Flamm, 1988). Another innovation that has dominated the global economy to this day was also developed with the support of governments: the internet. Computer-based networks and the internet emerged in the 1960s, supported by the US, French and British governments. Initially, the wide-spread use of networks in the US depended mostly on public funds and resulted in a substantial domestic network by the late 1980s, driven by large number of domestic users. Therefore, innovation in the internet relied both on the contribution of the US government, as well as on the size of the US market (Fagerberg et al., 2006, p. 369).

Anti-trust policy in the United States in the decades after the war also contributed to the rise of large-scale industrial research in large US companies. In the years between 1945 and 1950, there were several high-level anti-trust related lawsuits, involving for instance General Motors, Du Pont, AT&T, and Alcoa. In the anti-trust climate that prevailed in this time period, it was difficult for large industrial companies to acquire technology by acquiring other firms (Mowery, 1994).One example of this is Du Pont, whose central laboratory and development department was not authorized to search for promising technology or companies to be acquired. Due to these restrictions, firms increasingly had to rely on the internal development of innovative products (Mowery, 1994).

Small technology companies in the United States were also supported by relatively weak formal protection of intellectual property between 1945 and 1980. For example, companies operating in microelectronics and computer hardware and software benefited from permissive IP regimes – including liberal licensing and cross-licensing policies - allowing for technology to diffuse, while reducing the risk of litigation to newly established start-ups in relation to inventions originating within established large firms (Bruland & Mowery, 2004, p. 12).

These changes in the R&D strategies of large firms and the experience of prominent anti-trust litigation reinforced the position of small emerging high-tech companies in the post-war period. In Western Europe and Japan, the commercialization of new technologies continued to be done mainly by large organizations in industries in the pharmaceutical, electrical engineering, and other industries. In contrast, in the United States, basic research establishments in academic and public institutions, as well as in the private sector, were springboards for individuals who would leave these institutions to establish firms commercializing innovations based on newly acquired knowledge. This trend was supported by high levels of labor mobility and a relatively permissive legal climate facilitated the incubator role of universities and large firm. In retrospect, small new firms were responsible for a large number of significant innovations emerging in the United States after the war including semiconductors, computers, and biotechnology (Mowery, 1994).

The success of small technology firms in the US was also supported by the emergence of a sophisticated private financial system to support them during their infancy. Starting in the 1960s, the US venture capital market played an important role in the foundation and growth of new firms in different sectors including microelectronics, computers, and biotechnology, with annual flows of capital into industrial investments rising from about US $ 3 billion in 1969 to US $ 12 billion in 1983 and US $ 33 billion by 1989 (Mowery, 1994).

The following case study of the pharmaceutical and biotechnology industries in the United States illustrates the important role of public investment in R&D, as well as the increasing replacement of natural resource endowment by "created" resource endowment in the form of talent and human resources as important drivers of innovation-related success in the US in the postwar period.

Table 8: Historical case in point E: Innovation in Pharma & Biotechnology

Historical case in point E: *Innovation in Pharmaceuticals & Biotechnology*
While the US pharmaceutical industry had still been composed of several hundred small companies in the 1940s - with each being limited to a particular geographic region and accounting for 3 percent of the domestic market at most – by 1950, fifteen firms had emerged to define the American pharmaceutical industry.
Throughout the war, pharmaceutical companies increasingly relied on formal, in-house research and increasingly established cooperation agreements with US universities, in the expectation of achieving further breakthroughs. Biomedical research also received substantial support from the US government: according to the Pharmaceutical Manufacturers Association, in 1965, state funding contributed to almost two-thirds of total funding for biomedical research ("Pharmaceutical Manufacturers Association USA," 2015). Although funding by the National Institute of Health (NIH) has increased since this date, provide funding has increased even more. In 2014, the NIH invested about US $ 30 billion in medical research, while the US pharmaceutical industry invested about US $ 50 billion in R&D ("Pharmaceutical Manufacturers Association USA," 2015). The transformation of the pharmaceutical and biotechnology industries following the Second World War illustrates several important points about the evolution of innovation in this period.
On the one hand, while large-scale, industrial R&D remained important, many of the new drug discoveries in the "modern" pharmaceuticals industry resulted from the rise of biotechnology and new forms of cooperation, bringing together large and well-known pharmaceuticals companies with smaller innovative biotechnology firms.

On the other hand, the evolution of the pharmaceutical industry is representative of a fundamental shift in the nature of the US innovation context in the post-war decades. While innovation in the US during the Second Industrial Revolution – and in particular between 1900 and 1945 – had been driven to a large extent by the country's large natural resource endowment, in the post war decades, it increasingly relied on a non-physical, human resource "endowment" of engineers and scientists of American and foreign origin; this development was further supported by the significant inflow of skilled immigrants during and after the Second World War. While the size of the US market remained important, particularly in the internet and computer technology sector, post war America's endowment in specialized human capital, large compared to other industrial economies, provided a major stimulation to domestic innovation.

The above case illustration of the pharmaceutical industry in the United States is representative of a larger transformation process of drivers of innovation that has occurred in the decades since the Second World War. In this context, scholars of economic history have argued that such a transition from "natural" to "created" factor endowment (Bruland & Mowery, 2004, p. 15) was not limited to the United States. As Abramowitz (Abramovitz, 1994) and others have shown, other governments around the world have been able to erode some of these advantages, by investing in education, training and domestic R&D capabilities, making physical differences in natural resource endowment less critical in knowledge-driven industries. Furthermore, the post-1945 return to more open international trade and capital flows has enabled smaller countries to attain economies of scale by exporting their products, increasing the level cross-border flows of technology and knowhow. A more open trading environment also allowed for the import and adaptation of foreign technologies and combination with broad institutional change, this can help explain the rapid economic (catch-up) growth of countries like Singapore, Taiwan and South Korea, as the following section outlines in greater detail.

There are a number of contributions on the European experience of economic catch-up with Britain before World War One. Most notably, Thorstein Veblen and Alexander Gerschenkron offer two different perspectives on how less developed countries can catch up with technological and economic leaders. While Veblen highlights the possibility for late-coming economic developers to transfer and assimilate technology through market based mechanisms (Thorstein, 1915), Gerschenkron takes a more interventionist approach, arguing that institutions need to be created where prerequisites are missing, using the examples of banks in the case of Germany, and the role of the state in Russia (Gerschenkron, 1962). The Asian experience of economic, innovation and technological catch-up can be seen as a result of both of these perspectives.

Although some observers have attempted to classify the experience of Asian economies post World War Two as a Veblen-type, market based catch-up story, an abundance of recent literature shows that the catch-up strategies applied resemble more the Gerschenkronian, interventionist scheme (Fagerberg & Godinho, 2005). The following sections outline the experience of Japan, as well as Hong Kong, Singapore, South Korea and Taiwan. The case of Japan is chosen, as it illustrates the influence of government on technological and economic leadership in an Asian, postwar context.

Hong Kong, Singapore, Singapore and Taiwan – also known as the "Four Asian Tigers" – are chosen as examples of Asian economies that were influenced by the Japanese experience and subsequently also experienced rapid growth. Besides having gone through a phase of rapid growth, these countries share the experience of substantial and comparable structural reforms and economic policies, which helped them develop into major global producers (and exporters) in technologically progressive industries such as high-value electronics.

The experience of rapid economic growth and the underlying innovation and technology catch-up process in *Japan* after the Second World War needs to be understood in the context of its starting point, the so-called Meiji restauration of 1868. At the time, a party within the ruling elite established a new regime, with the explicit goal of strengthening the economy and the military strength of the state, which at the time had come under significant pressure by Western imperialism (Beasley & Beasley, 1995). Under the mantra of "A rich society and a strong army" (Fagerberg & Godinho, 2005, p. 518), the government actively intervened in order to make up for missing prerequisites for modernization: it reformed the legal system, physical infrastructure, and the education system, and supported the development of companies in what were considered as strategic industries. Much of the initiative to enable interaction between private and public actors came from the so-balled Zaibatsus, large and conglomerate businesses owned by families, and occurred in cooperation with government as well as the military (Fagerberg & Godinho, 2005, p. 518). These business groups may be seen as "institutional instruments" that replaced missing national institutions conducive to technological and economic development (Shin, 2013). R&D also greatly expanded in the time period leading up to the First World War, driven to a large extent by the needs of the military.

After the defeat of Japan in the Second World War, the role of the military and former Zaibatsu owners greatly weakened. Instead, the Japanese central government took on a leading role in driving technological development, in particular through the Ministry for Trade and Industry, also known as MITI. As the economy grew stronger, the role of private businesses increased, in particular through new business groups (called "Keiretsus"), which were based on the former Zaibatsus, which were liquidated by the US occupation u following the war. In this way, the Japanese economy rapidly gained productivity and competitiveness in selected industries including (in earlier years) the steel and shipbuilding industries, and subsequently in the automotive and (consumer) electronics industries.

During the period of rapid economic growth in Japan following World War Two, innovation in Japan consisted to a small degree of product innovation (e.g. product adaptations to specific market demand), but to a large degree of process innovations. In particular, organizational innovations in Japan's automotive industry such as the Kanban – also known as the "just-in-time" concept – allowed companies to exploit economies of scale and to be flexible at the same time and led to higher levels of through-put, efficiency in managing inventories ("zero inventory"), higher levels of quality ("total quality control"), as well as allowing Japanese firms to adapt quickly to the requirements of end-users, making the Japanese automotive industry unrivalled in its efficiency by the late 1980s (Fagerberg & Godinho, 2005, p. 519).

In *Hong Kong, Singapore, South Korea,* and *Taiwan*, Japan's successful postwar economic development and its underlying policies and practices generated a large degree of interest as a possible blueprint for their own future development. In retrospect, most scholars now agree that like in Japan, national governments in these countries appear to have played a very important role in the process of driving technological leadership and economic prosperity (e.g. Aoki, Kim, & Okuno-Fujiwara, 1997; Huff, 1995; Johnson, 1987).

In the 1960s, the governments of Taiwan and South Korea became heavily involved in the promotion of domestic economic development, following an export-led growth model that included measures such as tariff protection and the support of domestic industries that were considered to be of strategic importance, focusing at first on light manufacturing (e.g. textiles and electrical equipment), heavy manufacturing and chemicals in the 1970s, and automobiles in the 1980s. In Singapore, its government relied heavily on inward Foreign Direct Investment (FDI) in its industrialization efforts by using targeted FDI policies (Lall, 2000). More information about the emergence of Singapore as a global hub for innovation is provided below, in the section labeled as "Historical Case in point F: Singapore as an emerging innovation hub".

In Singapore, South Korea as well as Taiwan, an export-led growth level supported by the government was an important foundation of success. Furthermore, from the beginning, governments prioritized the expansion of education (particularly in engineering) and implemented policies supporting R&D and innovation, e.g. by providing generous funding for university-based research institutions (Lall, 2000).

Despite a number of similarities, there are also considerable differences in the experience of the "Four Tigers". For example, the four countries have taken different approaches in terms of their export policies. Hong Kong and Singapore, based on their small market size, had relatively open and liberal trade regimes encouraging free trade to allow for economies of scale. South Korea and Taiwan's approach was somewhat different and focused particularly on the protection and promotion of its export industries.

In South Korea, large, diversified business groups (so called chaebols) share many characteristics of the Japanese Keiretsus (former Zaibatsus) outlined earlier, by providing a corporate network that allowed for a more long-term approach to the development of new and risky technologies, due to available cross-funding within its conglomerate-like structure.

In Singapore, foreign multinational companies have played a more important role. In Taiwan's industries, small and medium-sized private firms are most prevalent (Fagerberg & Godinho, 2005). Due to their relatively small market size (with the relative exception of South Korea), industrialization in these economies was less geared towards the home market than in Japan, the United States and Germany previously. The reduction in trade barriers after the Second World War, and the deregulation of capital flows and financial markets in the 1980s, resulted in greater opportunities for international exports, as well as external finance (e.g. through foreign direct investment or lending) for these economies. In South Korea, economic development also depended to a great degree on foreign lending. While this provided important funds for economic development, it also made the economy more vulnerable, as the experience of the financial crisis in Korea towards the end of the 1990s illustrates. The following historical case in point provides a snapshot of policies that supported the emergence of Singapore as a leading technological innovation hub.

Table 9: Historical case in point F: Singapore as an emerging innovation hub

Historical case in point F: *Singapore as an emerging innovation hub*

Since its independence in 1965, Singapore has undergone a highly successful phase of technological and economic development. This success has resulted from an evolving national innovation system focusing on the attraction and leverage of MNCs and the resulting import of key technology, as well as the development of infrastructure and human resources to absorb and exploit new technologies rapidly (Edquist & Hommen, 2008, p. 71). Management practices that are traditionally encountered both in businesses and government in Singapore – openness, pragmatism, competitiveness and paternalism – have certainly contributed to the high level of efficiency and competitiveness of the Singaporean economy (Chong, 1987). Singapore's government supported development process since 1965 can be divided into four phases, following Edquist and Hommen (2008, p. 73):

1. *"Industrial take-off"* (1965 to mid-1970s): technology transfer through MNCs; emergence of Singapore as an offshore manufacturing hub in South-East Asia.

2. *"Local technological deepening"* (mid-1970s to late 1980s): improving domestic capabilities based on continued investment by MNCs in Singapore; related industries emerging in precision engineering and components assembly

3. *"Applied R&D expansion"* (from late 1980s to late 1990s): increasing applied R&D activities by global MNCs in Singapore; new public R&D institutions developing to support MNC product and process based innovation

4. *"Shift towards high-tech entrepreneurship and basic R&D"* (since late 1990s): emphasis on domestic technological innovation capability; emergence of high-tech start-ups; shift towards basic-level R&D; new science-based sectors, for instance life science

In retrospect, the decision to leverage foreign multinationals to stimulate domestic development resulted in a highly effective transformation process for Singapore. According to Edquist and Hommen (2008, p. 107), firms were able to increase competitiveness significantly by building up operational and adaptive competences, while having make less investments into R&D. However, this development strategy may also have slowed down the emergence of indigenous innovation in comparison to Taiwan and Korea (Edquist & Hommen, 2008, p. 107).

As a result, over the past 20 years, in line with the increasing sophistication of the domestic economy, Singapore is increasingly focusing on increasing indigenous levels of research and development. In a similar vain to China's government (albeit on a much smaller scale), Singapore's government continues to act as a 'developmental state', by implementing policies to improve R&D capabilities as an important element in Singapore's overall sustainable development strategy to make Singapore into a globally leading hub for innovation. Thus, the experience of Singapore in recent decades is illuminating as a further example of an active development state in the Asian context.

Thus, the government-led economic and educational policies outlined above clearly helped Japan, and later on Hong Kong, Singapore, South Korea, and Taiwan achieve rapid and prolonged economic growth. In the case of the Four Tigers, economic growth averaging more than seven percent for three decade and helped these countries achieve developed country status within this relatively short time period.

Lastly, the experience of interventionist government policies in East-Asian economies in the second half of the twentieth century also raises larger questions regarding the issue of policy measures in allowing for economic catch-up. Especially since the 1980s, there have been concerted efforts by the developed world, led by the United States and supported by several international organizations such as the World Bank, the IMF, and the WTO, to reduce the room of maneuver for such interventionist politics implemented by catching up countries. Some observers have interpreted this behavior as developed countries attempting to "kick away the ladder", which they previously climbed in order to achieve their economic and technological leadership positions (Chang, 2002).

5.1.4 Summary

In summary, the preceding discussion of the context of innovation evolving over time and across different industries and regions suggests that the evolving structure of innovation systems determine the development of innovation.

Innovation in the First Industrial Revolution was marked by the important role of individual entrepreneurs, an orientation towards crafts such as textiles, and little automation, with knowledge in wood- and metalworking being especially valuable.

The innovation system during the Second Industrial Revolution from the late nineteenth century was characterized by the increasing importance of state-funded, university based or external contract-based research, making innovation increasingly a routine occurring in large and integrated companies. In this time period, the importance of a large and homogenous market was also highlighted, with particular reference to the United States.

While it is more difficult to "summarize" the innovation context during the period following the Second World War, the evidence suggests a continuing trend towards increased technological sophistication and away from physical to immaterial (e.g. human) and "created" resources including technology and the internet in particular. The role of science and large-scale industrial R&D, as well as the role of the state in providing funding and sometimes also demand for R&D are important features of the innovation environment from this period onwards. This institutionalization of innovation and new forms of collaboration are well illustrated by the biopharmaceutical industry in Germany and the United States. In particular, the major financial support of large military as well as civil projects in the United States during the Cold War period clearly illustrates that innovation contexts differ and evolve not only based on sociocultural differences, but depending on the level of development and in coevolution with newly arising market needs. The following illustration summarizes the historical perspective.

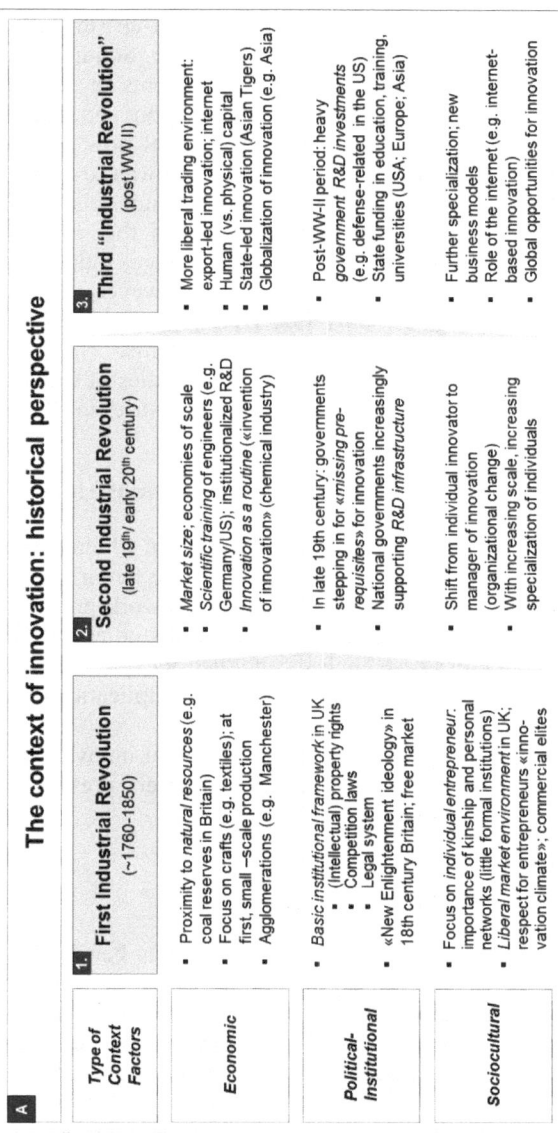

Figure 14: Summary of historical perspective

The experience of rapid economic and technological catch-up in Asian economies – in particular Japan, and later Hong Kong, Singapore, South Korea, and Taiwan – further illustrates the important role of governments in actively supporting innovation through investments into R&D, education and human resources, as well as the importance of a liberal trade and capital flow regime, which allowed also smaller economies such as Singapore and Taiwan to achieve rapid development through technology transfer and export-led industrial development. Thus, this historical perspective of innovation shows that comprehensive institutional change is necessary for innovation to emerge, rather than the development of individual "key" or "strategic" industries. However, as I also pointed out, due to the coevolving nature of innovation systems, individual firms or industries can make important contributions to innovation systems.

The following section outlines the evidence of four case studies of Chinese and non-Chinese companies operating in Strategic Emerging Industries in China.

5.2 Case studies: Innovation in China's Strategic Emerging Industries

This section presents the four case studies relating to innovation in China's Strategic Emerging Industries. The case studies are the result of the empirical, in-depth investigation of this research. The initial reference framework presented earlier provides the foundation for the presentation of the case studies and the empirical basis for the subsequent cross-case analysis and theory expansion. The new insights gained result in the theoretical and managerial implications and recommendations outlined in subsequent chapters.

Overview: The following section provides a brief overview of innovation in technology related industries in China. The subsequently chapters present the case studies on each of the four companies.

Innovation in technology related industries in China

Although China has for a long time been seen as the manufacturing bench of the world, its products were only rarely perceived as technologically cutting edge.

This is starting to change. Gradually, Chinese high-tech companies are becoming serious competitors to globally established corporations and are introducing new industry standards, for instance in telecommunications, mobile devices and online services. In order to improve their public perception and competitiveness at home and abroad, Chinese technology firms have been hiring western and western-trained Chinese managers, growing their overseas businesses with aggressive marketing campaigns that feature international athletes and Hollywood starts (Osawa & Mozur, 2014).

According to a number of high-level managers at Western and Chinese corporations, the technology sector in China is currently growing its expertise, talent and financial capability in a way that would significantly alter the structure of the global technology industry in the near future. While Chinese firms have long been seen as "fast-followers", a number of companies are starting to develop true innovation (Osawa & Mozur, 2014). To a large extent, this development is fuelled by increased investment in R&D. For example, Huawei Technologies based in Shenzhen (the second-largest telecommunications-equipment supplier by revenue in the world, after Sweden's Ericsson) increased its R&D expenditures per year by a factor of fourteen within one decade, from US$ 389 million in 2003 to US$ 5.46 billion in 2013 (Osawa & Mozur, 2014).

The shift towards more innovation among technology companies in China is also backed by the Chinese government, which seeks to transform the domestic economy from a global center of low-tech manufacturing to a major center of innovation (by the year 2020) and even to a global innovation leader by 2050. In figurative terms, it seeks to move from a model of "made in China" to "innovated in China", thus also reducing the country's dependence on foreign technology.

The following case studies reflect these recent changes in the innovation landscape of China: the shift towards higher domestic levels of innovation, as well the particular context of China that foreign and domestic companies face when operating in Chinese Strategic Emerging Industries in China.

The presentation of the case studies follows an overarching structure in order to increase the comparability and consistency:

- *Company profile*: a brief introduction to the firm's main businesses and industries, products, history, key figures, products and general strategy are provided including an overview of its organizational structure (e.g. main divisions and business units).
- *Firm context in China's Strategic Emerging Industries (SEI)*: this section outlines the situation of the firm in relation to the economic and political context of companies operating in China's Strategic Emerging Industries.
- *Firm-specific innovation practices*: this section provides specific examples of initiatives which are illustrative of innovation in China's SEI.

Summary and generalizability: this last section summarizes the findings and key insights and evaluates the generalizability of results.

5.2.1 Case 1: Bayer Material Science

Company profile

Bayer MaterialScience (abbreviated as BMS) is an independent subgroup within Bayer AG, a leading chemical and pharmaceutical company headquartered in Leverkusen, Germany. BMS emerged after Bayer AG underwent a major organizational restructuring in 2003 and 2004. The former business divisions were transformed into separate legal entities including Bayer Chemicals AG, Bayer CropScience AG, Bayer HealthCare AG and Bayer MaterialScience AG. Bayer Chemicals AG was renamed as Lanxess AG and was spun-off via an IPO in January 2005, while the other subgroups continue to be owned 100% by Bayer AG.

BMS is one of the world's largest manufacturers of polymers, which are core ingredients to high technology plastics used in industries such as automotive, construction, manufacturing, electronics, electrical, and sports and leisure. It comprises three product divisions - business units (BU):

- Polycarbonates (PCS)
- Polyurethanes (PUR)
- Coatings, Adhesives, Specialties (CAS)

Polycarbonate is a transparent, tough and heat resistant material with various applications for consumer devices and industry applications, e.g. for CDs, DVDs, eyeglass lenses, sports equipment, and headlights.

Polyurethanes (foam) are used, among others, as padding in furniture, as an isolating material for buildings, or as lightweight plastic for airplanes.

The *Coatings, Adhesives, and Specialties* business unit provides raw materials for coatings, adhesives and sealants, such as used in paints and glues (Meyer, 2013, p. 2). In 2014, the company had about 30 production sites around the globe, employing a workforce of 14,300 ("Bayer MaterialScience," 2013).

Its parents company, Bayer AG, was founded in 1863 as a dyestuff factory. By 1913, Bayer had become one of the largest German chemicals company, generating 80% of its revenues from exports and managing subsidiaries in Russia, France, Belgium, the United Kingdom and the USA. Key milestones in the development of the high-tech materials business have been the discovery of polyurethane in 1937 and polycarbonate in 1953. Polycarbonate became a popular plastic for household goods in the 1960s, and BMS led its introduction on optical storage devices, such as CDs, in the 1980s.

Recently, Bayer MaterialScience generated sales of EUR 11.2 billion (2013), slightly down from EUR 11.5 billion in 2012. As a lot of BMS's customers (e.g. in the automotive and construction industries) were hit by the recession starting in 2008, the company's sales also dropped from EUR 9.7 billion (2008) to EUR 7.5 billion (2009), recovering to EUR 10.2 billion (2010) and to EUR 11.5 in 2011 ("Bayer Annual Report," 2013).

Since the year 2000, the center of gravity of key customer industries, such as consumer electronics, has been moving to Asia, with leading manufacturers of mobile phones, computers and related devices being based in Japan, Korea and Taiwan, with China emerging as the main assembly base of the world for this type of products.

In 2013, Bayer MaterialScience spent €208 million (2012: €241 million) for research and development, accounting for roughly 6.5% of the Bayer Group's R&D expenses. According to the firm's annual report, the ratio of R&D expenses to sales in the subgroup itself was 1.9% (2012: 2.1%). Furthermore, Bayer MaterialScience spent €97 million (2012: €115 million) on joint development projects in cooperation with customers ("Bayer MaterialScience," 2013). As of 2013, 1,100 people were employed in research and development, mostly in the firm's Innovation Centers in Leverkusen, Germany, and Pittsburgh, Pennsylvania, United States, or the new facility for the Asia / Pacific region that opened in Shanghai, China, in 2013 ("Bayer MaterialScience," 2013). The innovation centers also provide R&D collaboration opportunities with customers and access to local talent and universities in three continents. According to BMS, the increased local presence in Asia is aimed at bringing research and development even closer to its growing customer base in the Emerging Markets and particularly in China ("Bayer Annual Report," 2013, p. 82).

The polycarbonate industry

The global polycarbonate industry is dominated by a small number of large players, of which Bayer is a market leader with an estimated 25 to 28% of the world market share (Meyer, 2013, p. 6). For many years, General Electric Plastics had been BMS' main competitor. In 2007 General Electric sold its plastic division to SABIC (a chemicals conglomerate from Saudi Arabia) which kept a dominant position. Other competitors are based in Japan or Korea, including the Japanese industrial conglomerates Mitsubishi, Teijin and Idemitsu, as well as Korean Samsung-owned Cheil.

The global polycarbonate industry has been marked by increased commodification of its products, with decreasing margins for standardized products. Innovation in this industry mainly takes place in creating new polycarbonate compounds that enhance its properties. On-going product development is aimed at e.g. creating thinner layers of polycarbonate that still retain core properties such as flame-retardancy. At the same time, some customer industries have been evolving fast as new designs created demands for new properties of plastics to be incorporated in the next generation of products (Meyer, 2013, p. 6).

Bayer MaterialScience operations in China

The demand for polymer materials is very strong in China, due to the rapid growth of the domestic manufacturing, automotive, and construction industries. Therefore, in 2006, Bayer opened an integrated polymers site in Shanghai Chemistry Industrial Park in Caojing, representing an expenditure of USD 1.8 billion through 2009, the firm's largest project ever outside of Germany (Bayer corporate website, 2014). Besides production, Bayer has invested in R&D facilities in China. The Bayer Polymer Research and Development Centre (PRDC), located in the Jinqiao Export Processing Zone in Pudong, Shanghai, was opened in November 2001 and further expanded in 2006 and 2013. The PRDC is an R&D platform focusing on the generation of new applications, materials and formulations.

In response to the shifting global geography of the industry, BMS moved the global headquarters of its polycarbonates business unit to Shanghai, China in July 2011 (Meyer, 2013). In addition, BMS expanded PRDC (see above) in Shanghai, China in 2013, complementing its two other global innovation centers in Leverkusen, Germany and Pittsburgh, Pennsylvania, United States. The leadership of Bayer Material Science has acknowledged that the greater China region will become the company's most dominant market by the end of 2015 (Bayer Corporate, 2015). The company expects annual sales in China to grow by an average of 11 percent over the next five years, from EUR 1.8 billion to roughly EUR 3 billion ("Bayer MaterialScience," 2013).

This suggests that BMS's increasing engagement in China is driven to a great extent by China's *large market size*. Besides increasing production capabilities in China, the interviews with R&D managers at the Polymer R&D Center (PRDC) in Shanghai confirmed the view China is increasingly seen by the company management as a place not only for production, but also for *generating innovation*.

In terms of the main motivations to conduct innovation in China, all interview partners individually confirmed the view that *proximity and access to local customers* was the most important single motivation of BMS to invest in its China operations. Due to the *dynamism and changes* in the market environment in China, being *close to the market and able to react swiftly* to changes was seen as indispensable prerequisites for success in the China market.

In terms of the *type of innovation* conducted in China, the respondents pointed out that currently, the Chinese innovation centers were mainly responsible for *adapting existing products to the needs of the Chinese market*, as well as for *finding new applications of existing products*, e.g. in the area of renewable energy. Based on interviews at BMS, as well as other companies in the industry, the China market lends itself as a location to further optimize existing products, e.g. by finding new applications.

In *comparing innovation activities* conducted in China as compared to other locations – in this case, compared to the headquarter location in Germany – it became apparent that BMS continues to conduct the prevailing part of its basic research in its headquarters in Germany. One reason for that is that more basic research needs a corresponding infrastructure, e.g. of supporting labs, that has to be built first. Also, proximity to customers is not that crucial as for application driven R&D. However, BMS develops its applications in the local market whenever possible, so that the process can be market-driven to *fit the local market needs*.

Further differences for BMS in conducting innovation that exist between China and Germany are related to *human resources*. On the one hand, based on the interviews conducted with BMS, as well as other companies in the industry, Chinese PhD graduates working in R&D laboratories tend to be *less independent* with research than foreign graduates. According to the respondents, this is mainly due to differences in the education system in China, which is less based on independent and creative thinking than the German system. Another factor is cultural: according to the respondents, in Chinese culture, there is traditionally a *greater emphasis on respecting older and/or more senior people*; open criticism of more senior colleagues is traditionally regarded as disrespectful in China. Therefore, in areas such as R&D where *idea generation and critical thinking* are important, both European as well as Chinese colleagues have had to go through an adjustment process to further improve the productivity of cross-national R&D teams working in China, in an effort to leverage cross-cultural particularities. However, the interviews also revealed the benefits of different work approaches between Chinese and western employees. As a general observation applying not specifically to BMS, while European and especially German R&D engineers tend to "over-engineer" products, focusing on quality perfection rather than marketability, their Chinese counterparts seem to follow a more *pragmatic, market-driven approach*, developing their own "smart ways of doing the job...[with a] mentality to try and test, like Americans" (R&D manager, personal communication, October 22, 2014).

The context of China's Strategic Emerging Industries (SEI)

Indeed, Chinese indigenous innovation policies support in particular those industries that are important customers for BMS, including the automotive, construction and electronics industries. Therefore, BMS – as do many other companies in China – may indirectly benefit from the SEI policies in terms of customer demand.

In addition, companies like BMS may benefit from government funding available for cooperation projects with local universities. As SEI policies are aimed at speeding up technology transfer from multinational corporations to domestic companies and research institutions, the Chinese government encourages such cooperation. In the words of one respondent: "The Chinese government sees our companies as a role model for Chinese companies, so local companies can develop and follow our example" (R&D manager, personal communication, October 22, 2014).

In projecting the *long-term impact* of SEI policies on foreign companies like BMS, respondents univocally declared that China's SEI policies will likely lead to domestic companies reaching similar levels of innovations as foreign companies. As Chinese companies have become more innovative in recent years, the Chinese policy environment has also evolved. According to the respondents, government policies have started to shift "from a drive from quantity of patents to quality" (R&D manager, personal communication, October 22, 2014), with more substantial examination for utility models and a shift towards (e.g. invention-based) quality patents.

Summary and generalizability

The experience of innovation at BMS in the context of China can be seen as a *success story* so far. Several factors that characterize the context of China were highlighted as being conducive to innovation: in particular, China's *large and dynamic market* provides a significant opportunity for BMS Innovation targeted at the China market needs to be adapted to local needs, e.g. by offering good value for money to customers who value "good enough" products more than additional features that they are not willing to pay for. This, in turn, stimulates pragmatic and innovative solutions that are close to the customers' needs. Therefore, the need to adapt to local demands in a competitive market environment such as China forces companies to find pragmatic solutions. In parallel China and Chinese customers strive for more innovative and sustainable solutions, e.g. for lightweight car parts or in renewable energy. Here BMS solutions can be used or developed based on the company's capabilities and globally interlinked R&D Centers' competencies.

Based on the respondents at BMS, as well as observations at other companies, the ability to serve the Chinese market is not only essential in tapping into the opportunities of this significant market. In addition, the context of China may also help BMS innovate successfully in other markets, e.g. in Latin America and Africa, where customer demand may be closer to the needs of current day China than in Europe, Japan or the United States. Therefore, for BMS, China is not only important due to its large market size, but also as a hub for global innovation in other world regions with comparable customer needs. The global headquarters' decision to locate one of only three global R&D centers in China provides further evidence of this.

Thus, the experience of BMS in China shows that foreign companies can benefit from innovating in China if they adapt to the local context, e.g. by identifying their local customers' needs.

5.2.2 Case 2: BYD- Daimler New Technology Co., Ltd.

China's automotive industry has experienced rapid growth in recent decades. In 2008, China became the largest automotive market in the world. Since 2009, its annual production of cars has exceeded that of the European Union or the United States. Despite a recent slowdown of the Chinese economy since 2014 and somewhat weaker demand for cars, new car sales in China are forecasted to represent 35 percent of the global auto market growth until 2020. As car penetration is expected to be only 15 percent by then, the remaining growth potential of the Chinese market is considerable. Similarly, by 2020, total sales in the Chinese market are estimated to be above 20 million per year (Wang, A; Liao & McKinsey & Company, 2012).

From the period of economic opening in China in the 1980s under Deng Xiaoping, the government has supported the development of the Chinese automotive industry as an important pillar of domestic development, due to the size of the industry, as well as its numerous linkages to other industries. Industrial policies have also sought to support the development of domestic automakers, e.g. by requiring foreign car manufacturers to form joint ventures with domestic manufactures and to share technology with their Chinese partners, in exchange for market access.

However, three decades after China implemented such policies, this strategy appears to be only partly successful. While policies have attracted investment and created around 30 millions of jobs (roughly 11 percent of the total workforce), it has done little to help the Chinese build strong brands, one of the original intentions. According to Liao Xionghui, vice president of Lifan Industry Group Co., a car and motorcycle manufacturer based in the southwestern municipality of Chongqing, "we have been trying to exchange market access for technology, but we have barely gotten hold of any key technologies in the past 30 years." This impression is also supported by industry data showing that foreign companies and their joint ventures continue to dominate the Chinese market ("China's Auto Joint Ventures Failing to Build Local Brands," 2012).

As the barriers are high for positioning in the markets for conventional gasoline-engine cars and realizable competitive advantages in the global competition still have to be identified, Chinese policymakers have seen a window of opportunity opening up for an early positioning in the electric vehicle market, where even established players are currently still at a comparatively early stage of development (Wang, A; Liao & McKinsey & Company, 2012). From 2004, the automobile industry policies have therefore been revised, focusing increasingly on technology and R&D. Therefore, the development of new mobility solutions – subsumed under the term "new energy automotive" – has been an important element in China's indigenous innovation strategy in Strategic Emerging Industries. Due to rapidly increasing urbanization in China, as well as environmental pressures, the ambition to identifying and developing a leadership position in new mobility solutions has been high on the agenda of Chinese policymakers.

This is also reflected in recent changes in policies, which have shifted towards the support of electric vehicle development in China. In order to achieve this goal, the Chinese central government spent RMB 100 billion (about EUR 11 billion) to support the domestic electric vehicle industry. Through these policies, the government hopes that electrical vehicle sales will rise to 500,000 units sold in 2015, and up to five million vehicles by 2020 (van Someren & van Someren-Wang, 2013, p. 205). Besides funding for R&D purposes and other industry investments, national governments also matter for innovation related to electric vehicles with regard to public vehicle infrastructure, as well as subsidy programs to make purchases of electric vehicles more attractive.

BYD and Daimler operations in China

The following case clearly illustrates this evolution in China's innovation environment. It is based on the experience of BYD-Daimler New Technology, founded in 2010 as a joint venture between the Shenzhen based company *Build Your Dreams* (commonly referred to as *BYD*), and *Daimler AG,* a leading German multinational automotive company. The joint venture launched the new brand *Denza* in 2012, specializing in the development of electric cars. The following section provides a brief introduction of Daimler AG and its activities in China, of BYD, as well as of their joint venture.

Daimler AG, founded in 1926 as Daimler-Benz and headquartered in Stuttgart, Germany, is the thirteenth-largest car manufacturer and second-largest truck manufacturer in the world by unit sales. Together with BMW and Audi, the two leading luxury car manufacturers (with Audi being the largest premium brand in China), Daimler AG controls about 80 percent of global luxury car sales (Geiger, 2015). Daimler AG's products also include buses and financial services. As of 2014, Daimler's main brands are Mercedes-Benz, Mercedes-AMG, and Smart Automobile. In 2013, Daimler AG had about 275,000 employees worldwide and achieved revenues of EUR 114.3 billion (Daimler AG, 2013).

In 2013, its automotive brand Mercedes Benz sold 15 percent of its vehicles in China, up 15 percent from the previous year. In addition, in 2013, Daimler AG opened a new R&D center in Beijing, investing RMB 865 million (about EUR 112 million) and employing around 500 engineers and designers at the new Mercedes-Benz's R&D facility. Furthermore, Daimler AG has also opened its first car engine factory outside Germany in China, which exclusively devoted to supplying local vehicle assembly plants (Daimler AG, 2013).

The *BYD Company Ltd.* ("BYD") is a Chinese manufacturer of automobiles and rechargeable batteries based in Shenzhen, Guangdong province. BYD (an acronym of "Build Your Dreams") has two major subsidiaries, BYD Automobile and BYD Electronic. In 1995, Chinese businessman Wang Chuanfu founded BYD for the production of rechargeable nickel cadmium batteries, in an effort to compete in the Chinese market against Japanese battery imports. By July 2002, the company had become the world's largest manufacturer, producing 65 percent of the world's nickel cadmium batteries and had a market share of more than 50 percent in the global mobile phone battery market, listing on the Hong Kong Stock Exchange in July 2012 (Wang & Kimble, 2010, p. 79).

In 2003, the parent company BYD founded BYD Auto Co., Ltd, following the acquisition of Tsinchuan Automobile Company in 2002. BYD Auto is involved in designing, developing, manufacturing and selling cars and buses under its label. According to the latest available information, in 2010, BYD Auto produced about 700,000 cars per year and sold a total of about 520,000 units that year, placing it as the sixth-largest Chinese automaker by vehicles sold in that year. By 2012, it produced over 600,000 vehicles, moving to 9th place in China's rapidly growing car market ("BYD Corporate Homepage," 2015). As mentioned earlier, in 2010, it entered into a 50:50 joint venture with Daimler AG for the development of an all-electric, new energy vehicle (NEV). The result of this cooperation, the *Denza*, was introduced to the public at the Auto China Motor

Show in Beijing in April 2012, and put on the market in December, 2014 ("BYD Corporate Homepage," 2015).

BYD-Daimler in the context of China's Strategic Emerging Industries (SEI)

The case of BYD and Daimler in China is a revealing one in relation to China as a context for innovation. This joint venture brings together two companies that have ambitious plans for the China market. While Daimler seeks to continue its rapid expansion in the Chinese market, BYD wants to position itself as China's (and later also as a global) leading manufacturer of electric vehicles by 2020, benefiting from global efforts to increase levels of sustainability, preferential treatment by China's policymakers, and its already existing capabilities in manufacturing rechargeable batteries. In the future, BYD seeks to offer further complementary products, becoming the first fully vertically integrated provider of electrical transportation with solar power, storage, and grid infrastructure (van Someren & van Someren-Wang, 2013, p. 206).

The case of BYD-Daimler illustrates that the particular context of China clearly offers a number of advantages to innovators. The size and dynamism of the Chinese market, as well as Chinese government policies to support the development of "New Energy Vehicles" (NEV) have created incentives for foreign and domestic companies to develop vehicles in China. However, all foreign companies are required to develop NEV in cooperation with local partners. The partnership agreements also require Chinese partner companies to "master key technologies" as a result of the joint venture in at least one of the three key technologies related to electric vehicles: battery; electrical engine; and vehicle control system (Director, Regulatory Affairs and Intellectual Property, Daimler Greater China Ltd., personal communication, 10 September, 2014).

In addition, Chinese innovation policies have defined key performance indicators (KPI) that determine which skills local development teams need to acquire as part of the cooperation, e.g. the ability to repair or service newly developed energy vehicles. According to one interview partner, these requirements, which can be subsumed under the formula "technology in exchange for market access", have been a constant for foreign automakers in the innovation context of China. In his view, "this mindset of technology transfer has also created administrative burden and makes it more difficult to do business in China" (Director, Regulatory Affairs and Intellectual Property, Daimler Greater China Ltd., personal communication, 10 September, 2014). China's increasing emphasis on up-stream technology and, in the case of electric vehicle, on vehicle software can be seen as a result from previous industrial policies that were only partly successful, as outlined above. However, according to the interview partners, these *policies are not likely to impose a threat* on Daimler AG in the future, as the company has taken a number of (operational, legal and human resources related) precautionary measures to avoid excessive technology spillover effects (Development and *Denza* project leader, BYD-Daimler, personal communication, 3 December, 2014).

Besides its *large and growing market* and the *support of policymakers* that seek to increase indigenous innovation in new vehicle technologies, interviews revealed that *human resources* are another important factor in the Chinese context of innovation, in two major ways. First, China has overtaken the United States as the number one country of origin of engineering graduates as well as PhD students in general. For companies like Daimler and BYD, the *size of the potential talent pool* for research and development activities has become an important reason for companies to innovate in China, even as wages in China have increased dramatically, reducing prior cost advantages of employing Chinese locals as compared to Western engineers.

In addition, respondents pointed out that they often have to spend more resources on training Chinese as compared to Western graduates, due to *less emphasis on applied and innovative teaching methods at Chinese universities.* Second, the interview partners pointed out that in terms of *cultural aspects*, e.g. affecting work attitudes, they perceived their Chinese engineers working in R&D as "very innovative when it comes to use-based innovation and hungry for new things". In the words of one respondent, "Chinese want to solve problems and are more goal-oriented – they are not as technology savvy and develop innovations for the love of technology " (Director, China Insights and Concepts, Daimler Greater China Ltd., personal communication, 18 November, 2014). While this point certainly provides a generalization, it points towards previous findings of cultural differences.

Chinese engineers were described as having a more market driven approach to innovation compared to their (mostly German) colleagues in the German headquarters, who often are focusing on "the next big thing" instead of sometimes small but practical innovations (Director, China Insights and Concepts, Daimler Greater China Ltd., personal Communication, 18 November, 2014).

Therefore, the current context of China is perceived by managers operating in China – of both Chinese and foreign origin – as a suitable environment to develop *pragmatic, user-centered solutions.* One respondent described the relationship between German and Chinese engineers in the conduct of innovation as "complementary", predicting that cultural differences that lead to different approaches to innovation between China and other countries would persist "for at least another twenty years" (Daimler Greater China Ltd., personal communication, 18 November, 2014).

Summary and generalizability

The example of BYD-Daimler illustrates several important points regarding innovation in the context of Strategic Emerging Industries in China. First, the *role of government* in the automotive industry continues to be pervasive. In the electric vehicle industry, national governments have significant influence, e.g. by providing *infrastructure, subsidies*, as well as *investments funds for R&D* in technology areas with high risks such as in electric vehicle technology. Although the Chinese government is taking a bet in diverting significant funds into this technology, *future payoffs for the industry* may be substantial. The role of *China as a source of human capital* was highlighted, both in its *quantitative as well as qualitative* (e.g. cultural) dimensions.

Lastly, the case of BYD-Daimler also illustrates the *risks* associated with the active intervention of the Chinese government in the new energy and particular electric vehicle industry. As SEI policies seem to be primarily targeted at increasing innovation among domestic players, e.g. though investment requirements that increase the level of technology transfer, foreign automakers may experience *increased competition by Chinese competitors* in the future. However, the case of Daimler AG shows that there is a growing awareness on the part of foreign companies of the *potential impact of SEI policies in the short and long run*, and that measures are being taken to address these issues.

5.2.3   Case 3: Haier Group

The speed and dimension of China's economic transformation since 1978 are unique in history. While China was one of the poorest countries in the world in 1978 (with per capita GDP at only one-fortieth of the US level), it has reached almost one-fifth of the US levels by now. However, in recent years, scholars as well as practitioners have wondered whether, and how, China can redirect its economic model away from its current focus on export-led capital investment towards consumption, efficiency, productivity and higher levels of innovation. Furthermore, the question has arisen whether China is capable of producing global brands that are able to foster real domestic innovation without technology transfer from western multinationals. These questions are at the forefront of mainstream analysis on China's next stage of economic development.

The case study of Haier provides possible answers to these questions. Being a China-based firm with global outreach, Haier been delivering pragmatic and user-centric innovations especially rural customers, who continue to make up a large share of the population. Amongst its Chinese competitors, Haier has held a leadership position in innovation, by closely aligning its product portfolio to the requirements of its target customers ("Haier: A Chinese Company That Innovates," 2010).

Company profile

The Haier Group is a leading multinational white goods and home electric appliance manufacturer that engages in the research, development, production and sale of a wide variety of household electric appliances. Its headquarters are located in Qingdao, China and sells a broad range of goods including washing machines, refrigerators, air-conditioning units, cell phones, microwaves, and televisions (*Haier Annual Report*, 2013). Haier's global market share in white goods was the largest in 2013, representing 9.7 percent ("Haier ranks as N° 1 Global Major Appliances Brand for 5th consecutive year," 2014) of total sales.

In 1984, Zhang Ruimin, the founder and CEO of Haier to this day, started off the company by taking over a defunct refrigerator factory in Qingdao. China was beginning to open up for global markets and foreign companies had increasingly started to look for Chinese partners to enter the local market. In that year, the Chinese refrigerator company partnered with German refrigerator company Liebherr in a joint venture, with Liebherr providing technological expertise in exchange for market access. From the beginning, the newly founded company had a relentless focus on product quality, which had been severely lacking in previous years. This strategy was successful and by 1986, the company became profitable, with annual sales growth of 83 percent ("Haier Corporate History," 2014).

With the integration of coherent quality management, the company returned to profitability by 1986 and sales grew at an average annual rate of 83 percent during the subsequent years. The company also diversified its products beyond refrigerators and became a market leader in white goods, first in China and increasingly also in overseas markets ("Haier Corporate History," 2014). Nowadays, its research and development activities are global. While its central research institute opened in Qingdao, China in 1993, Haier now has global R&D centers in five countries: China, Japan, Australia, the United States and Europe (in Germany and Italy).

The Haier way of innovating

Haier as a Chinese company stands out among its competitors in particular due to its international expansion strategy, as well as its approach to innovation.

First, regarding its *international expansion strategy*, unlike most other Chinese companies, Haier has not used a stepping stone strategy, the latter of which refers to companies moving first into other emerging, less-developed market and subsequently into developed markets such as Europe or the United States. Instead, Haier CEO Zhang Ruimin formulated the company's ambitious overarching strategy with the words "only play chess with the masters" (Zhang, 2009). Haier's international expansion strategy is based on entering advanced markets first, establishing the Haier brand and then making inroads in developing markets. That is why, in 1990, Haier exported its first batch of refrigerators to Germany. It is also why, in 1999, Haier established a refrigerator factory in South Carolina. According to its CEO, Haier seeks to "learn and grow by studying the competition" (Zhang, 2009).

Second, regarding its *approach to innovation*, Haier has become known for its market-driven, pragmatic and consumer-focused mode of innovation. Haier's engineers are consistently developing products that are close to customers' needs. Examples of this include mini fridges for young customers and freezers with a slightly warmer compartment (to keep ice-cream soft). Haier has also found new niches in the market, for instance inexpensive fridges for wine that have been deprioritized by European and US competitors. More recently, Haier has been a pioneer in offering wireless chargers (Haier and higher, 2013).

One often-cited episode illustrates and sums up the Haier approach to innovation. About 15 years ago, when a farmer from China's rural Sichuan province dialed into Haier's call center complaining that his washing machine was breaking down, service technicians found the plumbing clogged with mud. Rural Chinese were using the Haier machines, meant to wash clothing, to clean sweet potatoes and other vegetables. Rather than educating the farmer in correctly using washing machines, the Haier employee reported the farmer's experience to the company headquarters. Following this incident, the company developed a new type of machines that could wash both textiles as well as potatoes. This version developed in 2009 made Haier the global leader in laundry devices in terms of sales ("Haier: A Chinese Company That Innovates," 2010). Since then, Haier washing machines sold in Sichuan have been labeled as "mainly for washing clothes, sweet potatoes and peanuts" (Palepu, Khanna, & Vargas, 2006, p. 63).

In 2014, Haier won the *Fast Company Most Innovative Company Award* for being among the top-ten most innovative companies in China, e.g. based on its recent decision to restructure the entire company to focus on constant innovation, removing significant layers of middle management. While the 80,000 Haier employees were previously organized according to traditional corporate structures such as manufacturing or sales, they are now organized into 2,000 "zi zhu jing ying ti" (ZZJYTs). These are self-managed teams performing many different roles and each being responsible for profit and loss, with individuals being paid based on performance. In this model, when they identify opportunities, employees can suggest new projects or services. Based on votes from other employees, as well as suppliers and customers, projects are evaluated and implemented if accepted, making the winning employee the team leader of the suggested project (Haier and higher, 2013).

According to Haier CEO Zhang, the goal is "a free market in talent, so the cream rises" (Haier and higher, 2013) and further that "our nimble and delayered organization structure should allow us to implement our innovation culture of "Everyone to be a Maker", and enable business unit become energetic and innovative" (*Haier Annual Report*, 2013, p. 19). The increase in global sales to about USD 32.8 billion in 2014 (a fourfold increase from 2000), as well as a six fold growth in pretax profits in the same time period suggest that Haier's relentless focus on constant corporate transformation and innovation seems to pay off ("Haier ranks as N° 1 Global Major Appliances Brand for 5th consecutive year," 2014) in innovation as well as financial performance.

The context of China's Strategic Emerging Industries (SEI)

Interviews were conducted with innovation managers at Haier's headquarters in Qingdao, as well as the Haier Open Innovation Center. In general, the interviews revealed that Chinese government policies to support the development of SEI "really affect Haier" and that Haier "needs to work with the support of the government…[as] nobody can work without the government in China" (Haier Open Innovation Manager, personal communication, 4 December, 2014).

Being a leading technology company from China with high levels of innovative capabilities and global competitiveness, Haier can be seen as a showcase example of how Chinese policymakers may envision the future of other domestic companies. According to one respondent, central government leaders have outlined their vision of emphasizing technology transfer from developed markets to China, as well as increasing innovativeness of domestic players. In order to put such overarching guidelines into practice, local policymakers in the province of Shandong (in which the headquarter location of Qingdao is located) provide support to local players like Haier. One respondent confirmed the existence of such local government funds to promote innovation at Haier, but when asked about the nature of such support schemes, preferred not to comment on how Haier currently benefits from Strategic Emerging Industry government funds.

However, previously, Haier has benefited from government support in the form of financial support through low interest loans to facilitate the company's internationalization activities. Furthermore, in 2010, it benefited from government subsidies amounting to more than 15 billion RMB (roughly $2.23 billion at the time) for the rural population to buy Haier's products. Over the first four months of 2010, rural consumers purchased 41.7 billion RMB in household appliances as a result of the subsidies, an increase of about 510 percent compared to the previous year, resulting in the company doubling its sales in the first quarter of 2010 ("Haier: A Chinese Company That Innovates," 2010).

Haier views government policies to support domestic innovation as largely beneficial for the company. However, respondents also pointed out some of the potential downsides of the Chinese innovation context for Haier. For example, while current strategic and innovation priorities are largely in line with policy-makers' guidelines, one respondent noted that "current policies encourage us...but we cannot be sure whether the government will support us in the future; we only know that cooperation with partners and technology transfers will be supported...we need to work with the support of the government; nobody can work without the government" (Haier Open Innovation Manager, personal communication, 4 December, 2014).

Thus, currently, there is uncertainty over the future of government policies to encourage innovation in China. This is exacerbated by the fact that this support is viewed as essential to Haier, e.g. in the forms of preferential *government lending* to drive international expansion, *research cooperation projects with leading universities*, or indirectly though *stimulation of demand* among key customers of Haier, e.g. in rural areas of China or in government-led development projects involving building and construction.

Outlining further characteristics of the Chinese market that are seen as affecting innovation, respondents univocally highlighted the role of China's large and dynamic market as a key success factor of Haier. This is not only due to opportunities arising for *high sales growth* and *economies of scale*. According to one respondent, in China, there is more room for experimentation with new products compared to mature markets in Europe or the United States. While customers in China expect companies to respond quickly to newly emerging needs, they are "more willing to experiment and try new things". Therefore, at Haier, "we launch a lot of things – some succeed, some don't...in the China market, there is more room for experimentation" (Haier Open Innovation Manager, personal communication, 4 December, 2014).

Compared to Europe, the size and dynamism of the Chinese market leads Haier's engineers to do less upfront research compared to Europe or the United States. As a consequence, according to one respondent, in China, development windows are generally shorter; due to the large size of the market, "you can still succeed and make money". In contrast, in Europe, "development time is longer, but success rate is also higher". One way in which Haier achieves its fast development of new products is through cooperation with external partners. As one respondent notes, Haier relies a lot on external resources: "In my team, we gather technology needs and we scout for external resources from abroad (main R&D markets). We then cooperate with external companies, partner with them to provide us with the desired innovation and often repeat our cooperation...so far, this works well for us. Haier is very fast-to-market" (Haier Open Innovation Manager, personal communication, 4 December 2014).

Summary and generalizability

The case of Haier reveals how companies can conduct innovation in China by adapting to the local context and making use of some unconventional approaches. While the initial literature review of this dissertation, as well as the case experience of foreign and Chinese companies, suggested that the cultural heritage of China might be more conducive to pragmatic and applied innovation, Haier clearly confirms this view. Its innovation approach is consistently aligned with China's particular innovation context: on the one hand, its products correspond to the needs of local (e.g. rural) consumers. On the other hand, short product development cycles mean that Haier can respond to the needs of a fast-moving and demanding customer environment. Furthermore, its strategy is aligned with government priorities. The company can therefore be seen as one that not only benefits from indigenous innovation policies to promote development in SEI, but also as one that integrates support in a constructive and forward-looking way, by being highly innovative, responsive, customer-centric and managed by a visionary company leader who supports individual talent and provides growth opportunities to his employees.

The case of Haier also shows that the kind of innovation that Haier is most known for generating – pragmatic, close-to-market solutions and reverse innovation – should be seen as an equally valuable innovation to premium manufacturers' innovation based on technological leadership. In its innovation practice, Haier primarily corresponds to the local context and needs of the Chinese market, which offers significant opportunities due to its size and dynamism.

The following case study of Siemens AG – one of Haier's competitors in China – puts the example of Haier into perspective, by showing how a large

European company that has been inactive in the China market for a long time conducts innovation in the context of SEI policies.

5.2.4 Case 4: Siemens AG

Company profile

Siemens AG, founded in 1847 and headquartered in Munich, Germany, is one of the world's largest multinational electrical engineering and electronics companies. Its product range covers the areas of "electrification, automation and digitalization." ("Siemens AG Corporate Website," 2015). It is a leading provider of technology for energy-efficiency, providing products and services in diverse areas including automation, wind energy, power transmissions, infrastructure, and medical image as well as lab diagnostic devices ("Siemens AG Corporate Website," 2015).

Siemens' businesses are bundled into nine divisions and healthcare as a separately managed business ("Siemens AG Corporate Website," 2015):

- Power and Gas
- Wind Power and Renewables
- Power Generation Services
- Energy Management
- Building Technologies
- Mobility
- Digital Factory
- Process Industries and Drives
- Financial Services
- Healthcare

In 2014, Siemens AG and its subsidiaries had global revenues of EUR 71.9 billion (down from EUR 73.4 billion in 2013) and a net income of EUR 5.5 billion, employing a total of 341,000 employees at the end of 2014, of which 227,000 were located outside of Germany.

Siemens operations in China

The engagement of Siemens in China started in the late nineteenth century, with Siemens exporting pointer telegraphs to China in 1872 and constructing China's first electric tram in Beijing in 1899. With its long-term presence in China, it has become one of the leading and most respected foreign industrial companies in China. In the fiscal year 2014, Siemens generated revenues of EUR 6.4 billion in China, employing more than 32,000 employees ("Siemens R&D in China," 2015). In 2014, Siemens invested EUR 4.1 billion into research & development, which amounts to 5.7 percent of revenues, with 28,800 employees working in R&D ("Siemens R&D in China," 2015).

According to Siemens, "China is an ideal place to develop world-class innovations...[due to] diverse market needs and customers who are willing to try new things" ("Siemens R&D in China," 2015). Accordingly, the company has been increasingly investing in R&D capabilities, making its R&D facilities in China one of the most important bases for Siemens. At the end of the Fiscal Year 2014, Siemens had more than 4,500 R&D researchers and engineers, a total of 20 R&D hubs and more than 10,000 active patents and patent applications in China ("Siemens R&D in China," 2015).

At Siemens, divisions, as well as the Corporate Technology (CT) department carry out R&D. The interviews for this study were conducted with R&D managers at Siemens' CT unit in Beijing. While Siemens' individual businesses focus their R&D efforts on the next generations of their products and solutions, the aim of CT is to be a strong innovation partner for operational units and to help secure the technology and innovation future of Siemens as a technology leader ("Siemens AG Corporate Website," 2015). In the last ten years, Corporate Technology significantly increased its resources in emerging markets such as China and India. Currently, it maintains three research centers in Beijing, Shanghai, and Bangalore to push product innovation targeted to the needs of emerging markets.

It also established a technology-to-business center in Shanghai, which develops new business ideas in cooperation with internal and external partners. The goal is to use local resources and knowledge effectively to develop new solutions rather than adapting Western high-end products to local markets. Therefore, Corporate Technology serves as a service provider for the Siemens business units. As customers, the business unit can contract Siemens Corporate Technology for development project. In 2009, Siemens Corporate Technology's budget amounted to about USD 380 million, which was composed of contracted projects of the business units (about 60 percent), corporate financing (31 percent), and external funding (9 percent) (Widenmayer, 2012, p. 125).

According to the interview partners in Beijing, Corporate Technology's focuses on locally developing products for the domestic market, in order to meet Chinese customers' demands. This also helps in leveraging China's advantages in developing innovative products that can be marketed globally. One example of technologies that relate to China's innovation environment is "S.M.A.R.T.

Innovation" (which stands for "Simple, Maintenance-friendly, Affordable, Reliable, and Timely-to-market"), developing demand-driven innovations that have the potential to disrupt their industries, locally and globally ("Siemens R&D in China," 2015).

As the Chinese market is large and diverse in terms of economic development, culture, its climate and resulting consumer needs, as of 2013, the company has run dedicated innovation centers (located in Wuxi and Wuhan) that focus on indigenous innovation, working on Chinese customer demand-driven assignments in cooperation with local governments. In cultivating its close relationships with Chinese central and local policymakers, Siemens has also cooperated with the Ministry of Education in China in providing support in scientific research, vocational training and other education-related projects with a large number of Chinese universities ("Siemens R&D in China," 2015).

The context of China's Strategic Emerging Industries (SEI)

Over the past years, Siemens' role as a large multinational company in China has evolved. From the beginning, management has sought close cooperation with Chinese policymakers, to achieve a "win-win partnership" (Siemens Corporate Technology Beijing, personal communication, 9 September, 2014). However, the focus of cooperation has evolved from bringing manufacturing investment and expertise to China towards increasing domestic capabilities in R&D and innovation. As one respondent notes, "Simply said: ten years ago, when talking to government officials here, the main question was "could you build a factors" – now, the question is "could you bring research and development to us?" (Head of Research, Siemens Corporate Technology Beijing, personal communication, 9 September 2014).

According to Siemens' Head of Research in Shanghai, the evolution of the China context of innovation can be divided into four main phases:

Phase 1: China as an extended workbench ("manufacturing for global")

Phase 2: Localizing global product portfolio in China ("global for local")

Phase 3: Focus on the local product portfolio ("local for local")

Phase 4: Local products for global markets ("local for global")

All interview at Siemens agreed despite some of the risks associated to the innovation environment in China – in particular related to the protection of intellectual property rights – they view China at large as a great opportunity for Siemens to develop innovation that is increasingly not anymore limited to the Chinese market, but also targeting other destinations. As an early mover into the Chinese market, Siemens has developed mechanisms early on to benefit from the advantages that the Chinese market offers for innovation, while minimizing associated risks.

Siemens views the economic, institutional, and cultural and social context of China as favorable, due in particular to the size and dynamism of the market, a political environment that is highly supportive to innovation, the large talent pool of engineers in China, as well as the cultural and social aspects of China, which enable managers to effectively manage their local R&D teams based on suitable management and incentive structures (Siemens Corporate Technology Beijing, personal communication, 9 September 2014 and 17 November, 2014).

In particular, Siemens' experience regarding the cultural and social aspects of the employing R&D engineers in the context of China deserves further explanation. While an important motivation for multinational companies like Siemens to move innovation related activities to China used to be lower labor costs of engineers, annual wage increases averaging 10-12 percent in China over the past years have brought wage levels for Chinese middle and upper management employees with engineering backgrounds to European levels or even above. Qualified junior engineers in R&D departments of multinationals in China today can expect to earn equivalent of about EUR 28-30,000 per year. More senior R&D engineers often achieve rapid salary increases to about EUR 70,000 per year. Comparable salary levels in Berlin, Germany are roughly EUR 45,000 for junior R&D engineers and EUR 80-90,000 for highly qualified senior R&D engineers. In addition, competition in China is high for engineers that fulfill the requirements for multinational companies (e.g. English language skills; social skills such as the ability to work in teams and communicate effectively), which suggests that wage levels are likely to continue to increase. Therefore, lower wages are no longer a sufficient incentive for multinational companies to conduct innovation in China.

Instead, the interview respondents noted that all multinational companies nowadays need to focus on what makes innovation in China not more cost-efficient, but better than in other locations. In particular, respondents pointed out the pragmatic work attitude of Chinese R&D engineers. In relation to international R&D teams, this attitude was often contrasted with the approach of German engineers. As one respondent notes, "our Chinese R&D engineers are willing to try 80 percent solutions, put them on the market and make adaptations quickly when needed" (Siemens Corporate Technology Beijing, personal communication, 9 September 2014).

As the pressure for innovation and the need to respond quickly to changes in the market is especially high in China, but also growing globally, Siemens'

China operations can provide best practices and a benchmark along this dimension, in order to increase the responsiveness of its operations also in other locations. Furthermore, the interviews revealed that knowledge of the local market is seen as key to success in the Chinese market. Therefore, Chinese engineers can leverage their cultural background in their R&D activities for Siemens and are well equipped to effectively address the needs and aspirations of Chinese customers. This is why those development teams at Siemens CT that are focusing on pragmatic, cost-efficient solutions for the Chinese and other emerging markets are mainly consist of Chinese R&D engineers (Widenmayer, 2012, p. 128).

There are already products that can be seen as the result of what can be seen as a Chinese approach to innovation at Siemens, as well as examples of "reverse innovation" at Siemens in China. For example, in 2012, Siemens introduced an X-ray device specifically targeting the increasing demand for digital radiography in China's domestic market, after only one and a half years needed to develop this technology, significantly less than the global average project duration in the healthcare industry ("Siemens R&D in China," 2015). Siemens also developed a simplified computer tomography (CT) scanner focused on functionality and basic design. Originally developed for emerging markets, the CT scanner has subsequently also been sold in price sensitive market segments in developed countries, such as the United States and Japan.

According to the respondents interviewed for this study, China's indigenous innovation policies for the development of SEI are largely beneficial for Siemens. For instance, the Chinese government provides a corporate tax reduction of 25 percent for companies that receive a "high-tech status". Qualifying companies are required to own domestic R&D, register a part of their IP in China, prioritize local products and divert a pre-determined share of profits to the (local) operating company. Chinese government policies have also encouraged patent applications. As competitors are filing more patents in areas adjacent to Siemens core businesses, this may limit space to maneuver for Siemens in the future. As a result, the company has reacted by increasing its patenting activities in order to retain its ability to branch out into new innovative areas.

Although one respondent noted that "Chinese government policies are increasingly having an effect on foreign multinationals in China", he noted that these also offer "new possibilities" and further that domestic firms also needed to fulfill related criteria (personal correspondence, 9 September, 2014). Indeed, for Siemens, Chinese government-driven initiatives to increase domestic levels of innovations have provided platforms for cooperation on numerous occasions, as many of the company's products and solutions closely correspond to the current needs of the Chinese economy. For example, Siemens offers a number of products that can help local policymakers in achieving their own key performance indicators set by central and provincial governments, e.g. by increasing energy efficiency, improving infrastructure or by assisting in the development of smart cities, which have emerged in several parts of China.

Another illustration of this *symbiotic relationship* with policymakers is the Siemens Steam Turbine Engineering Hub. This cooperation seeks to increase expertise in steam turbine technology in China, developing future-generation steam turbines to modernize large-scale power plants that are still powered by coal. This is in line with Chinese government efforts to increase energy efficiency and the reduction of CO_2 emissions, as manifested in the Chinese 12th Five-Year Plan (China Central Government, 2011).

In a similar vein, since 2013, the company has cooperated with Jiaotong University in Shanghai to cooperate in research on several areas including turbine design and processing technology. Furthermore, in its Industrial Turbo machinery subsidiary, Siemens engineers and manufactures turbines and compressors for petrochemicals and power generation, as well as wastewater treatment plants in China ("Siemens R&D in China," 2015).

Summary and generalizability

Siemens AG in China exemplifies a foreign multinational company that has understood how to integrate into the Chinese market and to localize operations in order to derive benefits from the economic, political and sociocultural context of China, while minimizing the risks associated with conducting innovation in the context of China. The contextual factors that were found to be most conducive to innovation in China for Siemens include *market size and dynamism; proximity to consumers; the pragmatic market-oriented attitude of Chinese R&D engineers,* as well as their *market-specific knowledge;* the ability to *constructively engage with Chinese policymakers* to establish a *symbiotic relationship* that benefits both partners; the ability to *leverage China innovation for global operations.*

As challenges, respondents named in particular issues related to IPR protection in China, as well as changes as well as insecurity about the policy environments in China. Respondents also noted that in the long-term, SEI policies may lead to increased competition by local competitors, therefore potentially resulting in long-term adverse effects.

5.2.5   Summary and cross-case analysis

The following illustration summarizes the main insights gained about the four case studies, comparing and contrasting the main insights gained regarding the relevant economic, political-institutional, and sociocultural context of the four case study companies in China. Further implications resulting from the cross-case analysis will be discussed in the following chapter 6, which synthesizes the previous findings.

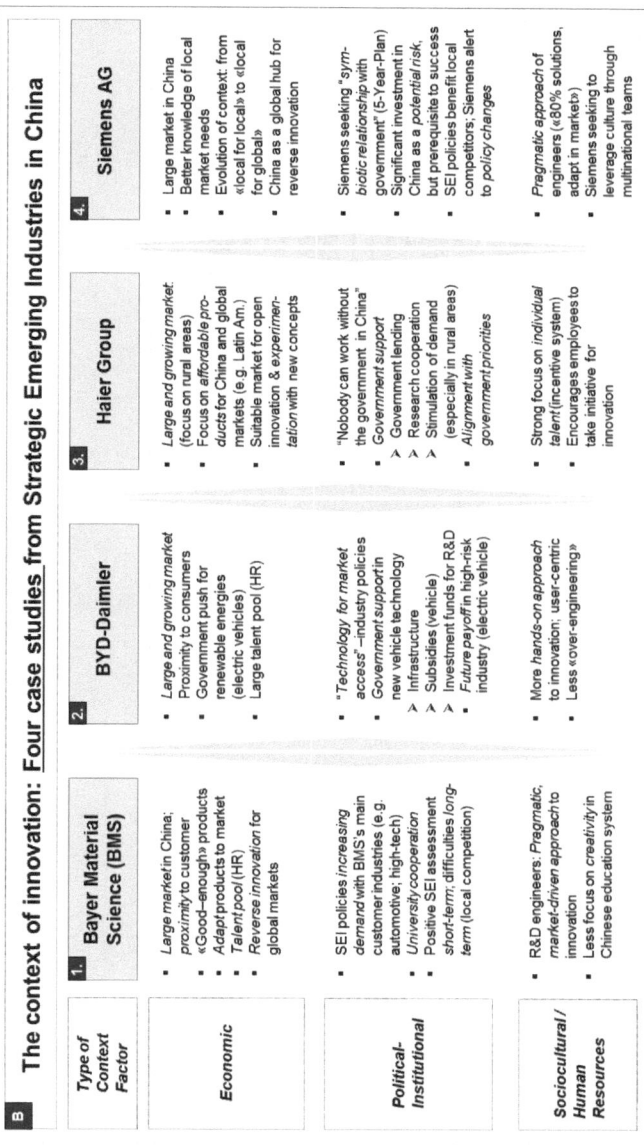

Figure 15: Summary of four case studies

6 Synthesis: integrating the historical perspective

Overview: Chapter 6.1 presents the findings about the economic, political-institutional, and sociocultural context of the four case studies presented, and places them in a historical perspective. Chapter 6.2 summarizes the previous points and results in a conceptualization of the analysis conducted, integrating the view on modern China with the historical perspective.

The historical analysis of the evolving innovation context, as well as the four case studies, resulted in new insights that are not covered in the initial reference framework presented in section 3.3. This chapter integrates the learnings based on the historical perspective as well as the four case studies. This is followed by a conceptualization of the economic, political-institutional, and sociocultural innovation context in China that integrates the main results of the historical and case-based analysis.

The initial framework presented in in section 3.3 is based on relevant existing literature on global R&D management, and the economic history perspective on innovation. The empirical data provide further insights that help in bridging these two literature streams, thus qualifying the initial framework. The following section syntheses the insights gained from the preceding analysis, regarding the economic, political-institutional, and sociocultural context of SEI in China. It considers each of the three main categories of context factors for innovation in SEI in China today, and re-evaluates them based on the insights from the historical perspective of innovation from the early Industrial Revolution to the present.

6.1 Synthesis of empirical results: economic context

The analysis of the four case studies shows that there are in particular four economic context factors that characterize the China market for companies in SEI today: the *size and dynamism of the market*; the significant potential for *further expansion*, e.g. in less developed (Western) regions; the potential of China as a *global hub* for reverse and increasingly also high-technology-based innovation; as well as a large and well-educated *talent pool*.

The *size of the Chinese market* stands out as the most important aspect. While there has been rapid development in the coastal regions of China, in the western, more inland regions, levels of income and development are significantly lower. For example, while the nominal GDP per capita in 2013 was US\$ 6,995 in China on average, it was US\$ 15,051 in Beijing, US\$ 14,547 in Shanghai, but only US\$ 6,003 in Xinjiang province in the northeastern part of the country (sharing borders with Russia, Mongolia, Kazakhstan, Pakistan, and India, amongst others), and US\$ 3,701 in the poorest province, Guizhou (Statistical Communiqué of the provinces on National Economic and Social Development, 2013). For the companies analyzed, this division provides challenge, as well as an opportunity.

On the one hand, wages in manufacturing, as well as in R&D have increased rapidly in recent years, which is a challenge for companies, as they can no longer achieve significant cost advantages from conducting R&D in China. However, as R&D units depend to a great degree on the ability to attract high-quality talent, and the latter are often located (or prefer to be located) in coastal business centers such as Shanghai, both foreign as well as domestic companies have suffered from increasing wage levels, while realizing that they cannot easily move R&D activities to more inland regions, as infrastructure there is lacking and highly-qualified, English-speaking talent can be hard to find and is not willing to move to those regions.

On the other hand, the current economic division of Mainland China between prosperous eastern regions and more inland regions also has potential benefits. While high disposable incomes and a large number of business customers in the coastal cities positively impact sales potential, lower wage levels in inland provinces means that companies can (and have) moved labor-intensive activities to those regions, thus continuing to benefit from cost savings and a relatively well-educated and disciplined work force. For R&D as a source of innovation, lower wages used to be an important driver for firms to locate R&D in China, but more recently, this benefit is waning, while the size of the Chinese market continues to be a major motivation.

From a historical analysis, market size is a well-documented driver of innovation, as it increases the potential payoff for the innovator. The most obvious parallel to China in this context is the United States. As outlined in section 5.1.2, in particular during the Second Industrial Revolution in the late 19th and early 20th century, the size of the US market was a major motivation for entrepreneurs to develop innovation in diverse sectors such as transport and communications. Due to the size and increasing integration of the market, success in innovation resulted in significant financial rewards. Important parallels can be drawn between the United States during the late 19th century and China today: the notion of "westward expansion" – referring today mostly to companies moving to "new frontiers" in the United States in the 19th century, could set a similar impulse for innovation and development in China in the near future.

Quite similarly, this strategy also includes political and economic risk. In the debate about the westward expansion of the United States in the 19th century, critics have often pointed out the unethical treatment of Native Americans in particular during the westwards expansion of the US railway network. In this period, a large number of Indian tribes were expelled from their traditional homelands and were forced by the US government to live in certain areas called Indian Reservations. In 1830, the Indian Removal Act under President Jackson provided funds and the authority to remove Indians by force if necessary. Similarly, in recent years, China has been criticized for its plans to industrialize western regions that often feature large ethnic minority populations. A notable example of this is Xinjiang – officially Xinjiang Uyghur Autonomous Region – which has a large Uyghur Muslim population. This comparison does not suggest that the unethical treatment of humans can be justified. Instead, it illustrates that there are reoccurring issues in the history of innovation, which should be solved based on now-available technological knowledge, as well as experience and social capital.

While market size still seems to matter significantly for innovation, the evolution of innovation has resulted in a number of changes in this context. While natural resources (e.g. coal and wood reserves) played a significant role for innovation in the early nineteenth century, its role has decreased significantly. As the discussion in chapter 5.1.3 has shown, significant improvements in the transportation and communications infrastructure since the Second World War has reduced transport and transaction costs significantly. Despite substantial reserves in a number of minerals such as graphite, aluminum, and zinc, China is not a natural resource rich country. Nevertheless, it has been able to develop rapidly, based on significant imports of important resources such as crude oil, natural gas and other commodities. Instead, as the discussion in section 5.1 suggests, in the 21st century, immaterial resources in the form of human capital have gained in importance.

Despite weaknesses in its education system, China's relatively well-educated workforce and large number of science majors have been significant drivers for innovation. The China market therefore embodies several features of the evolving innovation system outlined in section 5.1: it has a large market (similar to the United States); as its growing leadership in innovation has been occurring in the 20th and 21st century context, greater emphasis has been on education and training. In fact, many of the current SEI policies in China outlined in chapter three seek to improve scientific and practical training, in order to provide "resources" as a basis for better performance in indigenous innovation. The policies also point to another major driver of innovation, which was pointed out in a historical context and is clearly visible in China today: the role of government in spurring innovation.

6.2 Synthesis of empirical results: institutional & political context

Besides the size of the Chinese market, the experience of the four case study companies suggests that China's political and institutional framework has a significant impact on innovation. Chapter three outlined the specific policies that Chinese policymakers at the central and provincial level have introduced in order to increase the level of indigenous innovation, with a particular focus on Strategic Emerging Industries (SEI). While many of the respondents interviewed for this study view these policies as novel and are not sure about their motivation and (envisioned) ultimate outcomes, the historical perspective suggests otherwise.

Heavy government involvement in the support of innovation can be observed in several instances. For example, the German government in the 19th century provides a suitable example of a "development state" that heavily interfered in the development of key industries (e.g. the chemical industry) that were viewed as important pillars of broader economic development. Similarly, the United States interfered heavily in the development of key industries such as semiconductors, e.g. during the Cold War period. Therefore, the Chinese support of "pillar industries" or "Strategic Emerging Industries" per se should therefore not be interpreted as a particularly "Chinese" way of increasing innovative performance, or even as a uniquely Chinese part of an emerging "Beijing Consensus" (Ramo, 2004). Instead, the experience of active state involvement in development has a long tradition, even in liberal market economies such as the United States.

However, what differentiates China from previous historical examples is a greater knowledge of relevant parallel events in history, as well as the greater level of overall policy coordination in the Chinese system combining elements of socialist and free-market policies. Therefore, its combination of (market) size and stringent state involvement make China a unique case in history, as its government involvement spans across almost all areas of business activities, making it an important actor in all innovation-related activities across industries.

Furthermore, it is important to note that while the net effect of Chinese government investment in R&D related research and training seems to be positive, there are also challenges related to increased government funding. According to a personal interview with an editor of *China Daily*, China's largest English-speaking newspaper agency, within the Chinese government, there are significant concerns regarding misallocation and ineffective use of government funding for R&D-related projects. For example, the interview partner cited a case in which a group of scholars provided false documentation to apply for government funding, and was subsequently arrested for reasons of fraud. As the respondent notes, "you put in US$ 10, you get out US$ 1". (personal communication, senior editor at China Daily, in Beijing, November 17, 2014).

While this is a personal judgment rather than an official statement, it shows that the increased involvement of governments in the economy clearly also has potential drawbacks, e.g. related to corruption and inefficiency of fund allocation. This view has also been supported by the literature (Mahmood & Rufin, 2005). Therefore, despite the positive impact that government involvement in innovation can undoubtedly have, it is important to maintain a critical, fact-based perspective on this issue.

6.3 Synthesis of empirical results: sociocultural context

In the previous chapters, the importance of sociocultural factors for innovation was pointed out both with respect to individuals in the process of innovation (e.g. entrepreneurs), as well as collectively, related to the national culture.

The important role of the individual entrepreneur that would engage in a venture resulting in innovation was particularly pointed out for the early period of the Industrial Revolution. While the case of Boulton-Watt's steam engine points out the role of the inventing innovator, in subsequent decades, industries grew considerably, resulting in the needs for larger amounts of capital and investments. The example of Henry Ford illustrates this development towards large scale industries based on organizational and process innovation. From the late 19th century onwards, innovation increasingly resulted from corporate R&D departments, thus reducing the role of individual innovators.

China today increasingly follows this development. In line with increasing local corporate capabilities in R&D, innovation increasingly occurs in domestic as well as foreign firms, as the case studies of Chinese and foreign companies clearly indicate. In this process, China's sociocultural context is distinct from other countries. Based on the insights gained from the interview partners, cultural and social particularities of China can be observed also in its approach to innovation. Most notably, companies reported the pragmatic, market-driven mindset of R&D engineers as characteristic of the cultural aspect of innovation. As pointed out in chapter three, Confucian values related to respect for seniority, as well as China's education system also matter. From the perspective of Western managers, Chinese schools do not encourage creativity and innovative thinking as much as Western schools, making it difficult to operate in China without making organizational adjustments, e.g. by aligning employee incentives with innovative performance.

However, the analysis of companies like Siemens shows that foreign multinationals in particular may view China as an attractive location for R&D and innovation not despite, but rather because of these sociocultural differences. For example, the development of "good-enough", pragmatic product or business model innovations - for example in the form of reverse innovation - may emerge more easily in teams of Chinese engineers who may have a more pragmatic approach to innovation than their Western colleagues.

The historical perspective on sociocultural factors shows that innovation has emerged in different time periods and under different sociocultural paradigms. For example, in the context of Britain during the First Industrial Revolution, leading scholars have pointed out that the social recognition of entrepreneurship and business distinguished this society from other societies at the time such as France. This is to illustrate that innovation is always embedded in a sociocultural context, which may determine the level, quality and type of innovation that is most likely to occur.

Lastly, the case studies suggest that the evaluation of innovative performance needs to take into account the economic, political-institutional and sociocultural factors outlined above. As the current discussion about global R&D and innovation is still largely dominated by Western scholars and practitioners, a shift in mindset should take place, which identifies and recognizes these differences and re-evaluates previously-held conceptions and assumptions. A historical perspective can contribute to this task, by re-assessing concepts of "newness" and by providing a global perspective of innovation rooted in its particular context.

7 Theoretical and managerial implications

Overview: Chapter 7.1 provides the theoretical implications and resulting propositions of this study, based on the preceding historical analysis, as well as the synthesis (provided in the previous chapter) that re-evaluated innovation in current-day China from the historical perspective. In Chapter 7.2, the implications for management are outlined, based on the historical analysis and chapter 4 outlining the practical relevance of the present study.

This chapter first outlines the insights gained from the preceding analysis and provides theoretical implications. Using a theoretical perspective, this chapter also derives propositions extending current literature. Second, it presents managerial implications on how a historical perspective on innovation can increase managers' understanding of contemporary China and help them in making decisions resulting in long-term competitive advantage.

7.1 Theoretical implications

This dissertation contributes to the current literature on R&D management and innovation in China, by applying the insights of the historical perspective on innovation to current-day innovation in China, considering the economic, political-institutional and sociocultural context of innovation in Strategic Emerging Industries in China. The following theoretical implications can be drawn from the preceding analysis.

The study of innovation in a non-western context illustrates the western-centricity of innovation concepts, in which the Western propensity for the discovery of revolutionary new technology, as compared to the Asian inclination for modification, improvement and the application of technology, is often overemphasized, while both types of innovation are potentially necessary and equally lucrative, addressing different customer needs.

Increasing the robustness of literature on R&D management and innovation

The historical perspective also increases the *robustness* of the literature on R&D management in innovation in China. By placing the current perspective into a historical context, it provides a reference framework by showing under which conditions innovation emerged in the past, and how we can therefore evaluate China in the present. For example, Germany before the mid-nineteenth century was lacking the institutional foundations to support the emergence of industrialized, large scale R&D. In retrospect, the active involvement of the government through investments in training and research institutes made a significant contribution to the success of the German chemicals industry, with spillover effects to other industries. While the history of the German chemicals industry and the role of government has been covered by several previous contributions, in this dissertation, it serves as an example enabling scholars of innovation to gain an understanding of the underlying patterns of change that drive innovation. This results in the following proposition:

Proposition 1: The historical perspective on innovation in China increases the robustness of innovation literature, by providing a reference framework showing under which conditions innovation emerged in the past, and how we can therefore evaluate the phenomenon of innovation in China in the present.

Furthermore, research on R&D management and innovation focusing in China (and other emerging markets for that matter) has received increased attention only in recent years, based on institutional theory or corporate governance perspectives. In particular, previous contributions have tended to highlight differences between China and established markets, and in particular the novelty aspect of China's rapid evolution towards an innovation-intensive economy. A historical perspective helps in normalizing our perception of events in China, by showing that they are part of an evolving innovation system with parallels in history. Although the size of the Chinese market and population, as well as its political economy, make its transformation process unique in history, it still follows many of the predictions outlined by historians about late-developing countries. This provides new avenues of research, which may increase understanding about commonalities rather than focusing primarily on differences. In this way, scholars can assume intellectual leadership, by reconsidering established notions of innovation in new settings. This results in the following proposition:

Proposition 2: Building on previous contributions on R&D management and innovation in China that have highlighted in particular the differences between China and established centers of innovation, a historical perspective provides a fresh view that identify commonalities as much as differences and opens up new avenues of research that analyze relevant issues with this knowledge.

Establishing an identity for the study on innovation in China

Similarly, a historical perspective always relates an academic discipline to its own past and to other disciplines, and therefore helps in establishing an *identity* and a narrative for an academic discipline. For example, in "The Competitive Advantage of Nations", Michael Porter (1990) uses historical evidence of ten nations – for instance, "Patterns of National Competitive Advantage – The Early Postwar Winners" (chapter 7) - to identify nationals levels of productivity, supported by a suitable environment, as key to competitive advantage of nations. Similarly, a detailed historical perspective can provide a stronger identity and a more clear profile to innovation related research, by identifying the economic, political-institutional, as well as sociocultural factors that are paramount to achieving high levels of innovation. This study provides a first step into this direction and encourages further research to continue this task of using historical insights to sharpen the identity of research on innovation.

Proposition 3: The historical perspective on innovation provides an identity and a narrative for the study on innovation in China, by relating its core concepts to its past and to other disciplines.

Including the temporal element in the discussion on innovation in China

The historical perspective on innovation also informs our view on the management of innovation in current-day China, by showing that many of the aspects of the recent literature on R&D and innovation management in China can not only be related to differences in the economic, political and institutional, as well as cultural context of China. Importantly, the historical perspective on innovation occurring in evolving national and regional contexts adds an important temporal element to the discussion, by showing that China's current innovation context needs to be seen as constantly evolving towards a more mature level. Previous examples of government-supported modernization efforts in Europe, North America and more recent examples in Asian economics countries like Singapore suggest that governments can provide a suitable environment for innovation. The following proposition results:

Proposition 4: The historical perspective on innovation extends our perspective on innovation in China by adding a temporal element to the discussion, showing that China's current innovation context is constantly evolving towards more mature levels, as part of the evolution in structure of the innovations systems that influenced previous time periods.

Innovation in China: a development at different speeds

Lastly, the analysis of the innovation systems characterizing the First and Second Industrial Revolutions, as well as the post-war era, shows that its underlying structure has been changing over time. While innovation during the First Industrial Revolution was characterized by individual entrepreneurs with an orientation towards crafts such as textiles, the innovation system during the Second Industrial Revolution from the late nineteenth century was characterized by increasing standardized and routinized innovation occurring in large and integrated companies. The evidence of innovation occurring after the Second World War suggests a continuing shift towards increased technological sophistication and a shift away from physical to immaterial (e.g. human) and "created" resources including technology and the internet in particular.

With respect to innovation in China today, one can observe characteristics that resemble historical references from different time. Due to its enormous market size, as well as its ambition to achieve rapid transformation, the development of China towards an innovation-driven economy has been occurring at different speeds, depending on whether it is evaluated from an economic, political-institutional, or sociocultural point of view. For example, why the country has made enormous progress in terms of infrastructure economic prosperity, its political and institutional framework is still providing an obstacle to innovation, e.g. due to insufficient protection of intellectual property rights. However, as the case study analysis has shown, China has been making significant progress with respect to institutional quality. Thus, the historical perspective reminds us that economic development, and in particular national innovation systems, are complex processes that take time. Reference to history can therefore respond to observers that point out to existing insufficiencies in China's current innovation system.

Proposition 5: A historical perspective on innovation qualifies and extends our evaluation of current-day China as an emerging hub of innovation, as it illustrates that due to its size and complexity, China is developing as a hub for innovation at different speeds, suggesting that a differentiated assessment of China's innovation system is necessary, taking into account its economic, institutional and political, as well as sociocultural context.

Lastly, the historical analysis, as well as the four case studies have shown that firms are not only subject to the innovation system that they operate in, but they can also be important agents for change and help in shaping their respective innovation systems. The example of Werner von Siemens in the nineteenth century, as well of more recent efforts by foreign MNCs in China that have built alliances to leverage their position as technology leaders in China, show that this insight holds true in the past, as well as the present. Therefore, the following proposition can be made:

Proposition 6: While the historical evidence suggests that firms have been important agents of change in shaping their respective innovation systems, this should also hold true in the case of Chinese as well as foreign MNCs in China.

7.2 Managerial implications

Besides theoretical implications, this study also resulted in several implications for management.

The historical perspective developed in this study provides relevant insights regarding the issues that managers of innovation in China have cited in representative business surveys – such as the US-China Business Council survey from 2014, presented in chapter 4 – as well as during the interviews conducted for this dissertation.

According to the survey cited above, managers in China still experience important challenges in the China market, despite the significant opportunities that country's large domestic market offers. These include, but are not limited to, the increasing competition from Chinese competitors, an insufficient comprehension of China's institutional and sociocultural context; as well as issues related to human resources in China. The historical discussion of innovation evolving over time and in different regions, as well as the evidence from four Chinese and European companies, provides important insights that help in addressing such managerial issues. This chapter provides managerial implications that result from this study and the interviews, which can help in addressing these issues.

Lack of understanding of China's evolving innovation context

Most importantly, many foreign companies operating in China struggle to comprehend the rapidly evolving institutional sociocultural context of China and to manage innovation effectively in this dynamic environment marked by rapid changes. The experience of companies operating in Strategic Emerging Industries in China reveals that the decision to focus on innovation in China carries significant opportunities, but also risk. The SEI policies are aimed at increasing the level of indigenous innovation in China. As foreign multinational companies generally continue to have superior R&D and innovation capabilities compared to their Chinese competitors, from the perspective of foreign companies in China, this means that SEI policies seek to improve the competitiveness of their Chinese competitors, even if policies are supposed to also support foreign players.

However, in reality, based on representative surveys among foreign businesses in China, as well as based on personal interviews conducted in China for this dissertation, there is a great amount of concern among managers that China's industrial policies, in particular in SEI, are putting foreign companies at a disadvantage, e.g. through government procurement favoring domestic companies and low levels of transparency in the distribution of public funding for innovation (e.g. for public R&D projects involving local universities). The historical perspective provides important insights on this issue: case studies showing the evolution of previous industries and economies – for instance, Germany's chemical industry in the late nineteenth century, or semiconductors in the postwar United States - suggest that a broad arsenal of factors affecting innovation has existed throughout history and that they will continue to matter. It further qualifies popular managerial perceptions of "novelty" with respect to China's development, by showing that while China's development is unprecedented in scale, the underlying economic, political and social transformation processes are not.

The role of government in innovation

The historical analysis has shown that the involvement of national governments in economic development has been a reoccurring phenomenon in economic history, rather than a particular characteristic of China. As outlined by scholars of economic history, governments can – and have – been involved in supporting domestic development, to make up for missing prerequisites and to catch up with technological and economic leaders, as proposed by economic history scholars such as Alexander Gerschenkron and others. SEI policies can be classified in the same vein, as instruments to propel China to a higher level of development, as its growth model based on labor-intensive manufacturing and exports has come under pressure in recent years and has resulted in severe environmental damage, which makes a revised growth model based on technological innovation (e.g. in the areas of renewable energies) almost inevitable for the sustainable development of the country. Therefore, managers need to be aware of the enormous pressure that rests on Chinese policymakers to ensure continued growth of a country as large and geographically, ethnically and economically diverse country as China. Based on these insights on the evolving system of innovation applied to the current Chinese context, managers should be aware of the needs of China at its current development stage.

The coevolving nature of innovation in China

However, this does not mean that managers should accept every aspect of China's current innovation context. The historical analysis has shown that throughout different time periods, organizations have not only been subject to the business context they are operating in, but have also been actively shaping their operating environments. One example cited in this study is Werner von Siemens in the late 19th century, initiated the German Association for Patent Protection, and successfully represented his interests with the German government. This illustrates the coevolving nature of innovation, in which the emergence of innovation depends on institutional arrangements, but also shapes the latter. Applied to current-day China, this provides a possibility for organizations – both Chinese and foreign – to actively engage with political and other stakeholders, in order to become active subjects rather than objects in China's continuously evolving innovation context.

Based on the interviews conducted, in particular foreign companies in China are starting to realize this. As China's economic and legal institutions are maturing in line with increasing economic prosperity in the country, the domestic economy is becoming increasingly globally connected, as evidence such as increasing Chinese foreign direct investment shows. Large foreign multinational companies operating in China have become an integrated part of China's innovation system and a source of innovation and technology. While their integration (and investment) in China creates dependencies on the continued goodwill of Chinese policymakers, collectively, foreign multinationals also have significant bargaining power. Several of the companies interviewed therefore view the active engagement with Chinese policymakers, as well as industry alliances with competitors for market information as important aspects of their engagement in China. Foreign companies often implement such market-driven strategies to supplement other measures including market-specific legal and technology-related processes, product differentiation, as well as human resource related adaptations of established business practices.

Innovation and human resources in China

Furthermore, this study provides important implications for managers with respect to human resource issues in China, which can only be understood with a broader knowledge of historical relationships, as well as of China's particular context. The analysis has shown that innovation is the result of individuals, which places them in a central position in this process. As there are significant differences in culture between Chinese and most foreign (e.g. western) cultures, foreign companies need to be aware of these differences. Much of the related literature and human resource management concepts still assume a largely Western context and view different business environments as a case for adaptation to established practices.

The experience of companies in Siemens suggests that in order to achieve the largest benefit from innovating in China, companies need to develop a truly global and multicultural outlook. As R&D and innovation work require imagination, creativity and local knowledge, it is particularly important to realize the full potential of Chinese employees who bring this local knowledge to the table, as well as non-Chinese employees who may contribute with high levels of training and familiarity with established R&D processes.

From a historical perspective, the differences in the education system between China and European countries like Switzerland and Germany that were often cited by managers in China as obstacles to team-based innovation projects in China reflect the maturity of institutionalized innovation in the latter countries. While Germany made significant investments into university and research institute based R&D in the late 19th century (as outlined in chapter 5), China is currently in the process of developing such fundamental capabilities, as exemplified by the policies to increase indigenous innovation in Strategic Emerging Industries outlined earlier. The country has made significant progress in this respect, moving from a large, centrally planned and collectivist economy towards an increasingly economically open one, passing through different stages from technology acquisition over technology assimilation and more recently towards indigenous internal R&D and innovation.

From a managerial point of view, China's significant investment in R&D related education and training may result in significant improvements in employee skills in China in the mid- and long-term, with the potential to make human resources in China not a restraining factor for innovation (as it is

currently portrayed by the dominant literature) but increasingly an enabling factor.

Adjusting expectations: the time dimension of evolving innovation systems

Therefore, managers should be aware that this transformation process will take
time and that significant differences will continue to exist regarding institutional
arrangements, as well as resulting differences in the level of applied and practical
experience of graduates from the Chinese university system. Importantly, foreign
managers in China in particular should learn to distinguish between skills-based
and cultural differences, and provide the necessary support for the former, and
show respect to the latter.

Another issue that came up in almost all personal interviews with firms in
China – both Chinese and foreign – was related to retention rates of Chinese
employees in R&D departments in China, in which Chinese employees may
move to more lucrative positions more often than their foreign peers. In China's
urban areas, opportunities for well-educated young Chinese professionals are
abundant. For managers of multinational companies, high employee turnover
rates are damaging, as they increase the risk of intellectual property diffusion to
competitors and reduce the return on training and skills-based educational investments.

Once again, a historical perspective can be illuminating: based on the
evidence of previously less developed and now highly developed industrialized
countries, employees in less developed countries may value monetary reward
more highly in order to provide the basic needs of their families. This view has
been supported by previous studies (e.g. Fisher & Yuan, 1998) that show that
Chinese employees view wage levels, and opportunities for promotion and development as most important, followed by good working conditions and personal
loyalties from their superiors and the organization, while the level of interest in
the assigned work, as well as the wish for appreciation of work done is considered to be relatively less important compared to employees in the United States
and Western Europe. In the interviews, companies reported that some of
preferences have been changing; however, important differences remain, which
demand the attention of managers.

Integrating the cultural dimension of innovation in China

One important reason for this is culture: in China as a holistic culture, employees
seek recognition related to the goals and values of their teams or the orga-

nization. If (e.g. Western) managers fail to recognize this, they may not implement appropriate incentive systems for Chinese employees to stay in the organization. Therefore, incentive and rewards systems need to be tailored to the needs, disposition and inclinations of Chinese employees in order to secure the best talent as a crucial ingredient to success in innovation.

Localizing innovation in China

Some companies interviewed have shown that they are able to achieve this task: by defining their R&D centers in China as truly regional or local entities, they give sufficient independence and leeway to local human resource managers who understand the needs of their employees. These companies that have made these organizational adjustments to the local context typically report higher retention rates and lower turnover rates, as evidence from previously cited business surveys and interviews show.

Furthermore, firms may benefit from aligning their global innovation strategies with the comparative advantages of each innovation location, rather than focusing primarily on overcoming difficulties caused by differences across geographies. For example, given China's current stage in development, firms may currently focus on innovation activities targeting lower or middle-income customer segments in China, which also correspond to China-specific characteristics that were highlighted in the study, such as the ability of Chinese engineers to find pragmatic approaches that correspond to customers' needs.

Being a successful "glocal entrepreneur" in China's innovation system

As a result, from a firm perspective, developing a more global view on innovation building on regional (e.g. human resource related) comparative advantages can take away pressure from firms that struggle to implement globally-defined R&D management standards across regional and local subsidiaries. Rather than trying to realize a broad range of innovative activities in one location, MNCs with international innovation capabilities may consider focusing the development of specific types of innovation in China, for instance reverse innovation, as the case of Siemens demonstrates. Previous contributions have outlined such a "glocal" management approach that seeks to simultaneously integrate local as well as global strategic priorities (Robertson, 1995).

Aligning firm-based innovation with China's development needs

Given its current development stage, having passed through the stages of technology acquisition and assimilation, and currently moving towards the final stage of indigenous internal R&D and innovation, firms can "grow with China's development needs", and also benefit from the resources that are to increase domestic innovation, resulting in a mutually beneficial system of domestic and foreign companies, and policymakers. The policies in Strategic Emerging Industries outlined in this dissertation can potentially benefit domestic as well as foreign companies in China. The interviews showed that many foreign companies do not derive these benefits sufficiently due to a lack of understanding about China's overarching development priorities and related needs. For foreign firms in China, this means that China could serve as a global hub for certain types of innovation and gradually assume additional competencies, as its national innovation system further matures. Some firms have already started to act on this insight. Others have found themselves stuck in between, by trying to import foreign innovation processes to China, or by avoiding innovation-related activities in China altogether due to a lack of familiarity with local conditions.

Developing a long-term perspective on doing business in China

Despite continued risk and insecurities about China's future, the experience of other late-industrializing countries – as outlined in chapter 5 - suggests that even if China does not reach all of its ambitious objectives, the impressive level and comprehensiveness of national efforts to move R&D and innovation in China to a new step in development may still result in China's establishment as a base for innovation and scientific progress. Due to the large size of China and its importance in the region, this is almost certain to result in a new global world order of R&D and innovation. In the mid- and long-term, knowledge about these developments is thus essential for every globally minded business.

While it is impossible to cover all managerial decisions that may benefit from a broader historical perspective on innovation, this section has served to show that in general, this perspective can help managers become more mindful and integrative in their decision-making process, an important skill to have in an increasingly fast-moving, globalizing world.

8 Conclusion

Overview: Chapter 8.1 outlines the theoretical implications and summarizes the developed propositions. Chapter 8.2 recapitulates the managerial implications. Lastly, chapter 8.3 provides limitations of this research and further research opportunities.

Previous research on global R&D management and innovation has described the increasing importance of China as a hub for innovation and has called for a more fine-grained understanding of the context of China as a crucial condition for effective management of innovation in China. However, previous contributions have tended to neglect the historical dimension of innovation in China and the important implications of the historical perspective for theory and practice.

This study has attempted to provide a reference framework for managers in understanding and evaluating current developments as a consequence of past events and a precondition of the future. Reframing current approaches to management and identifying evolving schools of thought in management based on changing temporal paradigms, this can help managers deal with greater confidence with ambiguity and insecurity, which are frequently encountered in emerging markets such as China.

Based on the analysis and discussions of the previous chapters, this chapter summarizes the main findings of the preceding study, highlighting the central theoretical and managerial implications.

8.1 Summary: theoretical implications

This dissertation contributes to the current literature on R&D management and innovation in China, by applying the insights of the historical perspective on innovation to current-day innovation in China, considering the economic, political and institutional, as well as sociocultural context of innovation in Strategic Emerging Industries in China. The following theoretical implications were drawn from the preceding analysis.

First, it was shown that the use of historical analysis can increase the robustness of academic disciplines. In the case of innovation in China, it does so by providing a reference framework illustrating under which conditions innovation emerged in the past, and how we can therefore evaluate the phenomenon of innovation in present-day China.

Second, this dissertation extends previous contributions on R&D management and innovation in China that have highlighted the differences between China and established centers of innovation from different perspectives including institutional theory and corporate governance. Using the historical perspective, it seeks to provide a fresh view that identifies commonalities as much as differences and opens up new avenues of research that analyze relevant issues with this knowledge.

Third, it was shown that the historical perspective on innovation provides an identity and a narrative for the study on innovation in China, by relating its core concepts to its past and to other disciplines.

Fourth, the historical perspective on innovation extends our perspective on innovation in China by adding a temporal element to the discussion, showing that China's current innovation context is constantly evolving towards more mature levels, as part of the evolution in structure of the innovations systems that influenced previous time periods.

Lastly, in responding to the initial research question underlying this research, this dissertation suggests that a historical perspective on innovation qualifies and extends our evaluation of current-day China as an emerging hub of innovation, as it illustrates that due to its size and complexity, China is developing as a hub for innovation at different speeds, suggesting that a differentiated assessment of China's innovation system is necessary, taking into account its economic, institutional and political, as well as sociocultural context.

8.2 Summary: managerial implications

Besides theoretical implications, this study also resulted in several implications for management, which are summarized below.

First, the historical perspective provides a *broader understanding of China's evolving innovation context*, by showing the evolution of previous industries and economies, such as the chemical industry in Germany in the late 19th century, thus putting current events in China into a larger perspective. It also qualifies popular managerial perceptions of "novelty" with respect to China's development, by showing that while China's development is unprecedented in scale, the underlying economic, political and social transformation processes are not.

Second, this study has provided an extended understanding of the *role of government* in building up systems of innovation. Managers often struggle to grasp the role of government actors in China, for instance in Strategic Emerging Industries. The historical perspective has used evidence from several historical case studies to outline variations of government involvement and thus can increase understanding in the context of China.

Third, the historical analysis also illustrates the *coevolving nature* of innovation systems. It shows that firms are subjected to but are also actively shaping the innovation systems that they operate in. While current debates in R&D managements often focus on how foreign companies are affected by China's particular business environment, the historical perspective outlines different cases in which firms have actively shaped their innovation environments, showing not only *that* firms can shape their environments, but also *how* they can become actors in their innovation environments.

Fourth, the role of *human resources* was pointed out as a crucial element of innovation in China. Differences in approaches to R&D management need to be understood not only from a cultural, but also from a historical perspective. For example, different working styles and approaches to innovation also result from historically grown differences in the educational and training system. However, historical evidence suggests that investment in innovation-related education and training systems can have significant impact, as illustrated in the case of Germany in the 19th century. For managers, this means that in China, which is currently making significant investments into R&D related education and training, the effects of these investments may trickle through in the mid-term and long-term also with respect to innovation related skills of employees, with the potential to make human resources in China not a restraining factor for innovation (as it is currently portrayed) but increasingly an enabling factor.

Fifth, managers should learn how to adjust their expectations of China, by being aware that China's transformation process will take time and that significant differences will continue to exist regarding institutional. Foreign managers in China should acquire the ability to distinguish between skills-based and cultural differences, and provide the necessary support for the former, and show respect to the latter.

Sixth, it was shown that firms can benefit from aligning their global innovation strategies with the comparative advantages of China as an innovation location, rather than focusing primarily on overcoming difficulties caused by differences across geographies. Such a "glocal" management approach can also take away pressure from firms that struggle to implement globally defined R&D management standards across regional and local subsidiaries.

Seventh, the historical analysis and the company interviews stress the importance for firms to develop *symbiotic relationships with governments* in contexts with strong government presence. Rather than perceiving related institutional differences purely as organizational challenges, firms should learn how to "grow with China's development needs", resulting in a mutually beneficial system that includes the needs of businesses for new products and markets, as well as the needs of China's policymakers to increase indigenous levels of innovation.

Eighth, and finally, the historical perspective suggests that managers need to take a *long-term perspective in their assessment of China*. Despite continued risk and insecurities about China's future, the experience of other late-industrializing countries suggests that even if China does not reach all of its ambitious objectives, the level of national efforts to improve its R&D and innovation record may still result in China's establishment as a base for innovation and scientific progress and a shift in the global world order of R&D and innovation.

8.3 Limitations and further research

The following section outlines the limitations of the present research, which result in opportunities for future research in the field of R&D and innovation management.

Limitations of the present study

The findings of this study must be evaluated taking into consideration several limitations. However, these result in opportunities for future research.

Larger sample size: following the recommendations by Yin (1989) and Eisenhardt (1989), this dissertation chose four case studies spanning different industries and headquarter locations to provide new insights to theory and practice. While this sample size is sufficient according to the current literature on case study research, a larger sample size would possibly result in further variations and relevant issues, which could be analyzed from a historical perspective, which provides the main contribution of this study. For example, in the industries covered in this study, human resource related issues were of central concern to managers and served as a basis of analysis from a historical perspective. In other industries, different or additional issues may be important, which could be reevaluated from a historical perspective. Therefore, future studies may choose to take a historical perspective to analyze further industries and thus further extend our understanding of innovation in China based on historical analysis.

Choice of industry and company focus: this study focuses on innovation in China's Strategic Emerging Industries; out of seven different industries, this study focused on four companies in the areas of new materials (Bayer Material Science), premium engineering (Siemens AG), consumer electronics (Haier) and new automotive (BYD-Daimler). These industries were chosen as they offer a significant learning potential, representing important industries in China, which are affected by economic, institutional and political, as well as sociocultural context of China. In order to reflect the perspective on innovation in China from a foreign, as well as Chinese perspective, two of the four case studies were based on companies based in China. Similarly, interviews were conducted with both Chinese and foreign managers. While the choice of different industries and companies may have resulted in different findings at the firm level, from a historical perspective, the chosen industries represent the most significant aspects of innovation in China today and can serve as a basis for the historical analysis conducted.

Historical analysis: choice of time period and location

As this dissertation provides an evolutionary perspective on innovation, it is necessary to define a specific time period in order to derive meaningful conclusions, comparing different innovations evolving over time. Rather than focusing on specific key innovations only, this approach allows for the assessment of different economic, institutional and sociocultural aspects of innovation systems over time. The three subsequent time periods and regions analyzed in this dissertation – the First and Second Industrial Revolution, occurring mainly in Britain and later in continental Europe and the United States, as well as the time period after the Second World War, illustrated by examples from Europe, the United States, as well as Asia – were selected as they allow for a comparative analysis of innovation embedded in an institutional context, whose findings can be transferred and applied to the context of China. This selection should not downplay the importance of other time periods or regions. Instead, it opens up new opportunities for future research, which may include new regions.

Opportunities for future research

These limitations provide a basis for future research. In recent years, increasing contributions on R&D and innovation management in China has reflected the growing importance of China as an emerging global hub of innovation. In this context, there have been calls for new contributions that provide additional understanding of innovation in China. The historical perspective on innovation can provide a vehicle to better understand current issues in China, and to strengthen the robustness and identity of current research.

Future research could use evidence from additional time periods and regions to shed light on contemporary issues related to innovation. For instance, previous contributions have considered China's history of innovation, showing for instance that in imperial China, a lack of social recognition for wealth accumulation through entrepreneurship led to low levels of productive entrepreneurship and innovation. In exploring new research opportunities, it would be interesting to see which elements of China's long history of innovation still resonate today.

Future studies could also focus more on single issues related to innovation in China and relate them to a historical context. For example, while previous contributions have shown that China's approach to the protection of intellectual property rights and imitation may be different from a Western European one for cultural and historical reasons (as outlined in chapter 3), future studies could result in insights on how creativity – or in fact innovation itself – is perceived differently in China as compared to other countries, based on historical analysis.

Historical research in innovation continues to be carried out mostly by scholars outside of R&D management and innovation studies. In fact, over the past years, only little attention has been given to this part of the discipline. Most of the more recent contributions have come from business and economic historians such as David Landes, William Baumol, Joel Mokyr and others. While their work makes important contributions to the innovation literature, it does not write innovation literature. Reasons for the relative absence of historical research in innovation studies include a lack of appreciation of its importance and a lack of a method (Savitt, 1980, p. 52). While it is true to that historical studies tend to be lacking in methodological precision, the insights resulting from it can be substantial, providing thought leadership and a guide for further research in innovation studies.

In considering future opportunities for research, this study offers a rationale for the historical approach and a method, which can be applied in studies on innovation. It makes a step in this direction and encourages subsequent scholars to use the historical perspective to critically engage with current-day phenomena such as innovation in China. As the preceding analysis shows, this can result in important insights for theory as well as managerial practice. As China is likely to redefine a new global world order of R&D and innovation in the future, greater knowledge about the past and present can help managers in effectively addressing challenges related to China's particular context, providing a critical source of future competitiveness for MNCs in China.

References

2006 State Council, S. National Medium- and Long-Term Plan for the Development of Science and Technology (2006- 2020) (2006). Retrieved from http://www.most.gov.cn/eng/newsletters/2006/200611/t20061110_37960.htm

Abernathy, W. J., & Utterback, J. M. (1978). Patterns of industrial innovation. *Technology Review, 64*, 228–254.

Abramovitz, M. (1994). The origins of the postwar catch-up and convergence boom. *The Dynamics of Technology, Trade and Growth, Edward Elgar, Aldershot*, 21–52.

Acemoglu, D., Johnson, S., & Robinson, J. A. (2000). *The colonial origins of comparative development: An empirical investigation*. National bureau of economic research.

Acemoglu, D., & Linn, J. (2004). Market Size in Innovation: Theory and Evidence from the Pharmaceutical Industry. *Quaterly Journal of Economics, 119*(August), 1049–1090.

Adner, R. (2002). When are technologies disruptive? A demand-based view of the emergence of competition. *Strategic Management Journal, 23*(8), 667–688. Retrieved from http://doi.wiley.com/10.1002/smj.246

Altenburg, T., Schmitz, H., & Stamm, A. (2008). Breakthrough? China's and India's Transition from Production to Innovation. *World Development, 36*(2), 325–344. Retrieved from http://linkinghub.elsevier.com/retrieve/pii/S0305750X07002045

Aoki, M., Kim, H.-K., & Okuno-Fujiwara, M. (1997). *The Role of Government in East Asian Economic Development: Comparative Institutional Analysis: Comparative Institutional Analysis*. Oxford University Press.

Audretsch, D. B., Boente, W., & Tamvada, J. P. (2007). Religion and entrepreneurship. *Jena Economic Research Paper, 2007*(075).

Awokuse, T. O., & Yin, H. (2010). Intellectual property rights protection and the surge in FDI in China. *Journal of Comparative Economics, 38*(2), 217–224.

Ayres, R. U. (1990). Technological transformations and long waves: Part I. *Technological Forecasting and Social Change, 37*(1), 1–37.

Bantel, K. A., & Jackson, S. E. (1989). Top management and innovations in banking: does the composition of the top team make a difference? *Strategic Management Journal, 10*(1), 107–124.

Barton, D., Chen, Y., & Jin, A. (2013). Mapping china's middle class. *McKinsey Quarterly*, (January), 1–7. Retrieved from http://www.asia.udp.cl/Informes/2013/Mapping-Chinas-middle-class.pdf

Baumol, W. J. (1996). Entrepreneurship: Productive, unproductive, and destructive. *Journal of Business Venturing, 11*(1), 3–22.

Baumol, W. J. (2002). *The free-market innovation machine: Analyzing the growth miracle of capitalism*. Princeton University Press.

Bayer Annual Report. (2013). Retrieved November 15, 2014, from http://www.bayer.com/en/integrated-annual-reports.aspx

Bayer Corporate. (2015). Big player, big chances. Retrieved March 15, 2015, from htttp://www.materialscience.bayer.com/en/media/special/features/china.aspx

Bayer MaterialScience. (2013). Retrieved September 5, 2014, from http://www.bayer.com/en/integrated-annual-reports.aspx

BBC News. (2014, March 5). China congress reveals growth target and defence boost. *BBC News*. Retrieved from http://www.bbc.com/news/world-asia-china-26429481

BBC News. (2015). Sir Richard Arkwright. Retrieved January 15, 2015, from http://www.bbc.co.uk/history/historic_figures/arkwright_richard.shtml

Beasley, W. G., & Beasley, W. G. (1995). *The rise of modern Japan: political, economic and social change since 1850*. St. Martin's Press.

Beer, J. J. (1959). *The Emergence of the German Dye Industry*. Urbana: University of Illinois.

Benner, M. J., & Tushman, M. L. (2003). Exploitation, exploration, and process management: The productivity dilemma revisited. *Academy of Management Review, 28*(2), 238–256.

Berg, B. L., & Lune, H. (2004). *Qualitative research methods for the social sciences* (Vol. 5). Boston: Pearson .

Bessant, J., & Tidd, J. (2007). *Innovation and entrepreneurship*. West Sussex, England: John Wiley & Sons.

Black, T. R. (1999). *Doing quantitative research in the social sciences: An integrated approach to research design, measurement and statistics*. Sage.

Boer, H., & During, W. E. (2001). Innovation, what innovation? A comparison between product, process and organisational innovation. *International Journal of Technology Management, 22*(1), 83–107.

Boisot, M., & Child, J. (1996). From fiefs to clans and network capitalism: Explaining China's emerging economic order. *Administrative Science Quarterly, 41*(4), 600–628.

Bonoma, T. V. (1985). Case research in marketing: opportunities, problems, and a process. *Journal of Marketing Research, 22*(2), 199–208.

Bower, J. L., & Christensen, C. M. (1995). Disruptive technologies: catching the wave, (February), 43–54.

Breznitz, D., & Murphree, M. (2011). *Run of the red queen: Government, innovation, globalization, and economic growth in China*. Yale University Press.

Bromley, D. B. (1986). *The case-study method in psychology and related disciplines*. Chichester: Wiley .

Brown, J. S., & Hegel, J. (2005). Innovation blowback: Disruptive management practices from Asia. *The McKinsey Quarterly*, 35–45.

Bruche, G. (2009). The Emergence of China and India as New Competitors in MNCs' Innovation Networks. *Competition & Change, 13*(3), 267–288.

Bruland, K., & Mowery, D. C. (2004). Innovation through time (pp. 1–49).

Bruton, G. D., & Ahlstrom, D. (2003). An institutional view of China 's venture capital industry Explaining the differences between China and the West. *Journal of Business Venturing, 18*(2), 233–259.

Bruton, G. D., Ahlstrom, D., & Puky, T. (2009). Institutional differences and the development of entrepreneurial ventures: A comparison of the venture capital industries in Latin America and Asia. *Journal of International Business Studies, 40*(5), 762–778.

BYD Corporate Homepage. (2015). Retrieved January 15, 2015, from http://www.byd.com/index.html

Cain, P. J., & Hopkins, A. G. (1993). *British Imperialism: Crisis and Deconstruction, 1914-1990* (Vol. 2). Longman Group United Kingdom .

Carosso, V. P. (1987). *The Morgans: Private International Bankers, 1854-1913*. Harvard University Press.

Chandler, A. D. (1990). *Strategy and structure: Chapters in the history of the industrial enterprise* (Vol. 120). MIT press.

Chandler, A. D. (1992). Organizational Capabilities and the Economic History of the Industrial Enterprise. *Journal of Economic Perspectives, 6*(3), 79–100.

Chang, H.-J. (2002). *Kicking away the ladder: development strategy in historical perspective*. Anthem Press.

Cheung, G. C. K. (2009). *Intellectual property rights in China: politics of piracy, trade and protection*. Routledge.

Child, J., & Tse, D. K. (2001). China's transition and its implications for international business. *Journal of International Business Studies, 32*(1), 5–21.

China Central Government, C. C. G. China's 12th Five-Year Plan (2011). Retrieved from http://www.britishchamber.cn/content/chinas-twelfth-five-year-plan-2011-2015-full-english-version

China eyes new strategic industries to spur economy. (2012). Retrieved October 12, 2014, from http://www.reuters.com/article/2012/07/23/us-china-economy-strategic-idUS BRE86M03R20120723

China's Auto Joint Ventures Failing to Build Local Brands. (2012, August 22). *Bloomberg*. Retrieved from http://www.bloomberg.com/news/articles/2012-08-22/china-s-auto-joint-ventures-failing-to-build-local-brands

Chong, L.-C. (1987). History and managerial culture in Singapore: "Pragmatism", "openness" and "paternalism." *Asia Pacific Journal of Management, 4*(3), 133–143.

Christensen, C. M. (1997). *The Innovator's Dilemma. Business*. Harvard Business Review Press.

Clarke, D. C. (1991). What's Law Got to Do with It-Legal Institutions and Economic Reform in China. *UCLA Pacific Basin Law Journal, 10*(1), 1–76.

Cooper, J. R. (1998). A multidimensional approach to the adoption of innovation. *Management Decision, 36*(8), 493–502.

Corbin, J., & Strauss, A. (1994). *Grounded theory methodology. Handbook of qualitative research*. Sage London.

Crafts, N. (1977). Industrial Revolution in England and France: Some Thoughts on the Question,"Why was England First?" *The Economic History Review, 30*(3), 429–441.

Crafts, N. (2005). The first industrial revolution: Resolving the slow growth/rapid industrialization paradox. *Journal of the European Economic Association, 3*(2-3), 525–534.

Cremer, D. de. (2015). Understanding Trust, In China and the West. *Harvard Business Review*, (February 11, 2015).

Culture shock: Japanese firms in China. (2010). *The Economist*. Retrieved from http://www.economist.com/node/16542339

Daft, R. L. (1978). A dual-core model of organizational innovation. *Academy of Management Journal, 21*(2), 193–210.

Daimler AG. (2013). 2013 Annual report of Daimler AG. Retrieved October 15, 2014, from http://www.daimler.com/dccom/0-5-1591240-49-1591645-1-0-0-0-0-0-12591-7164-0-0-0-0-0-0-0.html

Damanpour, F. (1991). Organizational innovation: A meta-analysis of effects of determinants and moderators. *Academy of Management Journal, 34*(3), 555–590.

Damanpour, F., & Evan, W. M. (1984). Organizational innovation and performance: the problem of" organizational lag." *Administrative Science Quarterly, 29*(3), 392–409.

Danneels, E. (2004). Disruptive technology reconsidered: A critique and research agenda. *Journal of Product Innovation Management, 21*(4), 246–258.

Davidson, W. H. (1979). Factor endowment, innovation and international trade theory. *Kyklos, 32*(4), 764–774.

Demirbag, M., & Glaister, K. W. (2010). Factors determining offshore location choice for R&D projects: A comparative study of developed and emerging regions. *Journal of Management Studies, 47*(8), 1534–1560.

DiMaggio, P. J., & Powell, W. W. (1991). *The new institutionalism in organizational analysis* (Vol. 17). University of Chicago Press Chicago, IL.

Dwyer, S., Mesak, H., & Hsu, M. (2005). An exploratory examination of the influence of national culture on cross-national product diffusion. *Journal of International Marketing, 13*(2), 1–27.

Eayrs, J. (1971). *Diplomacy and its Discontents* (Vol. 107). Toronto: University of Toronto Press.

Eckermann, E. (2001). *World history of the automobile*. Society of Automative Engineers.

Edquist, C., & Hommen, L. (2008). *Small country innovation systems: globalization, change and policy in Asia and Europe*. Cornwall: Edward Elgar Publishing Limited. Retrieved from http://www.elgaronline.com/view/9781845425845.xml

Eisenhardt, K. M. (1989). Building Theories from Case Study Research. *Academy of Management Review, 14*(4), 532–550.

Eisenhardt, K. M., & Graebner, M. E. (2007). Theory building from cases: Opportunities and challenges. *Academy of Management Journal, 50*(1), 25–32.

Espinasse, F. (1877). *Lancashire Worthies*. Simpkin, Marshall & Company.

Eurostat. (2014). Eurostat Database. Retrieved March 15, 2015, from http://epp.eurostat.ec.europa.eu/portal/page/portal/eurostat/home/

Evan, W. M. (1966). Organizational lag. *Human Organization, 25*(1), 51–53.

Fagerberg, J., & Godinho, M. M. (2005). Innovation and catching-up. *The Oxford Handbook of Innovation. Oxford University Press, New York*, (March), 514–543. http://doi.org/10.1093/oxfordhb/9780199286805.003.0019

Fagerberg, J., Mowery, D. C., & Nelson, R. R. (2006). *The Oxford handbook of innovation*. Oxford Handbooks Online.

Financial Times Lexicon. (2014). Retrieved August 15, 2014, from http://lexicon.ft.com/Term?term=emerging-markets

Fisher, C. D., & Yuan, X. Y. (1998). What motivates employees? A comparison of US and Chinese responses. *The International Journal of Human Resource Management, 9*(3), 516–528.

Flamm, K. (1988). *Creating the computer: government, industry, and high technology.* Brookings Institution Press.

Flynn, D. M. (1985). Organizational and environmental effects on innovation: A comparison of two countries. *Asia Pacific Journal of Management, 2*(3), 150–163.

Foreign Policy. (2014). A New Definition of Chinese Patriotism. Retrieved September 11, 2014, from http://foreignpolicy.com/2014/09/11/a-new-definition-of-chinese-patriotism/?wp_login_redirect=0

Francis, D., & Bessant, J. (2005). Targeting innovation and implications for capability development. *Technovation, 25*(3), 171–183.

Freeman, C. (1995a). The 'National System of Innovation' in historical perspective, (March 1993), 5–24.

Freeman, C. (1995b). The "National System of Innovation"in historical perspective. *Cambridge Journal of Economics, 19*(March 1993), 5–24.

Fuchs, H. J., Kammerer, J., Ma, X., & Rehn, I. (2006). *Piraten, Fälscher und Kopierer: Strategien und Instrumente zum Schutz geistigen Eigentums in der Volksrepublik China. , .* Wiesbaden: Gabler.

Gadiesh, O., Leung, P., & Vestring, T. (2007). The battle for China's good-enough market. *Harvard Business Review, 85*(9), 80.

Garcia, R., & Calantone, R. (2001). A critical look at technological innovation typology and innovativenss terminology: a literature review. *The Journal of Product Innovation Management, 19*(2), 110–132.

Gassmann, O., Beckenbauer, A., & Friesike, S. (2012). *Profiting from innovation in China.* Springer.

Geiger, F. (2015, February 5). Daimler Forecasts 10% Profit Growth. *Wall Street Journal.*

Gerschenkron, A. (1962). *Economic backwardness in historical perspective.* New York: Frederick A. Praeger.

Giamartino, G. A., McDougall, P. P., & Bird, B. J. (1993). International entrepreneurship: The state of the field. *Entrepreneurship Theory and Practice, 18*, 37.

GLORAD R&D Database. (2014). GLORAD R&D Database. Retrieved from www.glorad.org

Golvers, N., & Verbiest, F. (2003). *Ferdinand Verbiest, SJ (1623-1688) and the Chinese heaven: the composition of the astronomical corpus, its diffusion and reception in the European republic of letters* (Vol. 12). Leuven University Press.

Gould, D. M., & Gruben, W. C. (1996). The role of intellectual property rights in economic growth. *Journal of Development Economics, 48*(2), 323–350.

Govindarajan, V., & Ramamurti, R. (2011). Reverse innovation, emerging markets, and global strategy. *Global Strategy Journal, 1*(3-4), 191–205.

Govindarajan, V., & Trimble, C. (2009). Is Reverse Innovation Like Disruptive Innovation? *Harvard Business Review, September*, 8–9.

Griffiths, T., Hunt, P. A., & O'Brien, P. K. (1992). Inventive activity in the British textile industry, 1700–1800. *The Journal of Economic History, 52*(4), 881–906.

Grossman, G. M., & Helpman, E. (1993). *Endogenous innovation in the theory of growth*. National Bureau of Economic Research.

Grueber, M., & Studt, T. (2014). *2014 Global R&D Funding Forecast*. Columbus, Ohio.

Habakkuk, H. J. (1962). *American and British technology in the nineteenth century: The search for labour-saving inventions*. University Press.

Haber, L. F. (1971). *The chemical industry, 1900-1930: international growth and technological change*. Clarendon Press Oxford.

Haier and higher. (2013). Haier and higher. *The Economist*, (12 October, 2013).

Haier Annual Report. (2013). Qingdao (China). Retrieved from http://www.google.de/url?sa=t&rct=j&q=&esrc=s&source=web&cd=4&ved=0CDoQFjAD&url=http%3A%2F%2Fwww.haier-elec.com.hk%2Fimage%2Fpdf%2Freport%2F20140520101417118.pdf&ei=DWLjVIGvHIHkUMyjg_AH&usg=AFQjCNGW1eS545vofX9NAMX0dxGHJRctLQ&bvm=bv.85970519,d.bGQ

Haier Corporate History. (2014). Retrieved September 15, 2014, from http://www.fundinguniverse.com/company-histories/haier-group-corporation-history/

Haier ranks as N° 1 Global Major Appliances Brand for 5th consecutive year. (2014). Retrieved November 15, 2014, from http://www.haier.com/uk/newspress/pressreleases/201401/t20140107_204053.shtml

Haier: A Chinese Company That Innovates. (2010). Retrieved October 15, 2014, from http://www.forbes.com/sites/china/2010/06/17/haier-a-chinese-company-that-innovates/

Han Chinese proportion in China's population drops: census data. (2011). Retrieved November 15, 2014, from http://news.xinhuanet.com/english2010/china/2011-04/28/c_13849933.htm

Hart, S. L., & Christensen, C. M. (2002). The Great Leap: Driving Innovation From the Base of the Pyramid. *MITSloan Management Review, 44*(1), 51–56.

Harzing, A.-W. (2006). Response Styles in Cross-national Survey Research A 26-country Study. *International Journal of Cross Cultural Management, 6*(2), 243–266.

He, Y., Tian, Z., & Chen, Y. (2007). Performance implications of nonmarket strategy in China. *Asia Pacific Journal of Management, 24*(2), 151–169.

Helpman, E. (1992). *Innovation, imitation, and intellectual property rights*. National Bureau of Economic Research.

Hicks, J. R. (1932). *The Theory of Wages*. London: MacMillan.

Hofstede, G. (1980). Culture and organizations. *International Studies of Management & Organization, 10*(4), 15–41.

Hofstede, G. (1991). *Culture's consequences: Software of the mind*. London/New York: McGrawHill.

Hofstede, G. (1993). Cultural constraints in management theories. *The Academy of Management Executive, 7*(1), 81–94.

Hofstede, G. (2001). *Culture's consequences: Comparing values, behaviors, institutions and organizations across nations*. Sage.

Hofstede, G. (2003). What is culture? A reply to Baskerville. *Accounting, Organizations and Society, 28*(7), 811–813.

Hofstede, G., & Bond, M. H. (1988). The Confucius connection: From cultural roots to economic growth. *Organizational Dynamics, 16*(4), 5–21.

Huang, X., Schroder, B., & Steffens, P. (1999). The Chinese steel industry in transition: industry perspective on innovation policy. *R&D Management, 29*(1), 17–25.

Huff, W. G. (1995). The developmental state, government, and Singapore's economic development since 1960. *World Development, 23*(8), 1421–1438.

Jin, J. (2005). *Technological Capability Generation in China's High-tech Industries: Experiences from China's Mobile Phone Industry.* University of St. Gallen. Retrieved from http://www1.unisg.ch/www/edis.nsf/SysLkpByIdentifier/3086/$ FILE/dis3086.pdf

Johnson, C. (1987). Political institutions and economic performance: the government-business relationship in Japan, South Korea, and Taiwan. *The Political Economy of the New Asian Industrialism, 136.*

Jones, G. (2013). *Entrepreneurs, Firms and Global Wealth since 1850* (No. 13). *Harvard Business School Working Paper.*

Jones, G., & Khanna, T. (2006). Bringing history (back) into international business. *Journal of International Business Studies, 37*(4), 453–468. http://doi.org/10.1057/palgrave.jibs.8400198

Keeley, L., Pikkel, R., Quinn, B., & Walters, H. (2013). *Ten Types of Innovation: The Discipline of Building Breakthroughs.* New Jersey: John Wiley & Sons.

Keupp, M. M., Friesike, S., & von Zedtwitz, M. (2012). How do foreign firms patent in emerging economies with weak appropriability regimes? Archetypes and motives. *Research Policy, 41*(8), 1422–1439. Retrieved from http://linkinghub.elsevier.com/retrieve/pii/S0048733312000807

Keupp, M. M., Palmié, M., & Gassmann, O. (2012). The Strategic Management of Innovation: A Systematic Review and Paths for Future Research. *International Journal of Management Reviews, 14*(4), 367–390. http://doi.org/10.1111/j.1468-2370.2011.00321.x

Knight, K. E. (1967). A descriptive model of the intra-firm innovation process. *Journal of Business, 40*(4), 478–496.

Knowles, H. (2004). *Why Management and Business Studies Need History.* Sydney.

Lai, E. L.-C. (1998). International intellectual property rights protection and the rate of product innovation. *Journal of Development Economics, 55*(1), 133–153.

Lall, S. (2000). Technological change and industrialization in the Asian newly industrializing economies: achievements and challenges. In L. Kim & R. R. Nelson (Eds.), *Technology, learning, & innovation: Experiences of newly industrializing economies* (1st ed., pp. 13–68). Cambridge: Cambridge University Press.

Landes, D. (2000). Culture makes almost all the difference. In L. E. Harrison & S. P. Huntington (Eds.), *Culture matters: how values shape human progress* (1st ed., pp. 2–13). New York: Basic Books .

Landes, D. (2003). *The unbound Prometheus: technological change and industrial development in Western Europe from 1750 to the present.* Cambridge University Press.

Lazonick, W. (2002). Innovative Enterprise and Historical Transformation. *Enterprise & Society, 3*(1), 3–47.

Lazonick, W. (2004). Innovative Enterprise and Historical Transformation. *Enterprise & Society, 3*(1), 3–47.

Lee, J. (2009). State owned enterprises in China: reviewing the evidence. *OECD Occasional Paper*, 6–7.

Lerner, J. (2000). *150 years of patent protection*. National bureau of economic research.

Li, Poppo, L., & Zhou, K. Z. (2008). Do managerial ties in China always produce value? Competition, uncertainty, and domestic vs. foreign firms. *Strategic Management Journal*, *29*(4), 383–400.

Li, Y., Liu, Y., & Zhao, Y. (2006). The role of market and entrepreneurship orientation and internal control in the new product development activities of Chinese firms. *Industrial Marketing Management*, *35*(3), 336–347.

Lu, Y., Tsang, E. W. K., & Peng, M. W. (2008). Knowledge management and innovation strategy in the Asia Pacific: Toward an institution-based view. *Asia Pacific Journal of Management*, *25*(3), 361–374.

Luecke, R. (2003). *Managing Creativity and Innovation. The Harvard business essentials series* (Vol. 12). http://doi.org/10.1108/14626000510612286

Lundvall, B.-A. (1992). *National innovation system: towards a theory of innovation and interactive learning*. London: Pinter.

Machiavelli, N. (1513). *The Prince. Italica* (Vol. 55).

MacLeod, C. (2002). *Inventing the industrial revolution: The English patent system, 1660-1800*. Cambridge, UK: Cambridge University Press.

Mahmood, I. P., & Rufin, C. (2005). Government's dilemma: The role of government in imitation and innovation. *Academy of Management Review*, *30*(2), 338–360.

Mahoney, J. T., & Chi, T. (2001). Business strategies in transition economies. *Academy of Management Review*, *26*(2), 311–313.

Mansfield, E. (1986). Patents and innovation: an empirical study. *Management Science*, *32*(2), 173–181.

Markides, C. (2006). Disruptive innovation: In need of better theory. *Journal of Product Innovation Management*, *23*(1), 19–25.

Marquis, S. S. (1923). *Henry Ford: an interpretation*. Detroit, Michigan: Wayne State University Press.

Massey, M. E. (1952). *Ersatz in the Confederacy* (1st ed.). South Carolina, Columbia: University of South Carolina Press.

McGregor, J. (2010). *China's drive for "indigenous innovation": A web of industrial policies. American Chamber of Commerce in China*.

Meyer, K. (2013). Bayer MaterialScience: Opportunities in Complex Global Value Chains. *CEIBS*.

Meyer-Thurow, G. (1982). The industrialization of invention: a case study from the German chemical industry. *Isis*, *73*(3), 363–381.

Miles, M. B. (1979). Qualitative Data as an Attractive Nuisance: The Problem of Analysis. *Administrative Science Quarterly*, *24*(4), 590–601.

Mintzberg, H. (2005). Developing theory about the development of theory. In K. G. Smith & M. A. Hitt (Eds.), *Great minds in management: The process of theory development* (pp. 355–372). Oxford: Oxford University Press.

Mokyr, J. (2002). *The gifts of Athena: Historical origins of the knowledge economy*. New Jersey: Princeton University Press.

Mokyr, J. (2010). *Entrepreneurship and the industrial revolution in britain. The Invention of Enterprise: Entrepreneurship from Ancient Mesopotamia to Modern Times.* Princeton University Press Princeton, NJ.

Mokyr, J. (2012). Entrepreneurship and the industrial revolution in britain. In D. S. Landes, J. Mokyr, & W. J. Baumol (Eds.), *The invention of enterprise: Entrepreneurship from ancient Mesopotamia to modern times* (pp. 183–211). New Jersey: Princeton University Press.

Mowery, D. C. (1981). *The emergence and growth of industrial research in American manufacturing, 1899-1945.* Stanford University.

Mowery, D. C. (1983). The relationship between intrafirm and contractual forms of industrial research in American manufacturing, 1900–1940. *Explorations in Economic History, 20*(4), 351–374.

Mowery, D. C. (1994). The US National Innovation System: Origins and Prospects for Change. In D. C. Mowery (Ed.), *Science and Technology Policy in Interdependent Economics* (pp. 79–106). Norwell, Massachusetts: Kluwer Academic Publishers.

Murmann, J. P. (2000). Knowledge and competitive advantage in the synthetic dye industry, 1850-1914: the coevolution of firms, technology, and national institutions in Great Britain, Germany, and the United States. *Enterprise and Society, 1*(4), 699–704.

Murmann, J. P., & Homburg, E. (2001). Comparing evolutionary dynamics across different national settings: the case of the synthetic dye industry, 1857–1914. *Journal of Evolutionary Economics, 11*(2), 177–205.

Neal, L. (1994). The finance of business during the Industrial Revolution. *The Economic History of Britain since, 1700*(2), 151–181.

Nelson, R. R. (1993). *National innovation systems: a comparative analysis.* New York: Oxford University Press.

Nicholas, T., Aghion, P., Howitt, P., Romer, P. M., Gilbert, R., & Shapiro, C. (2011). What Drives Innovation? *Antitrust Law Journal, 77*(3), 787.

North, D. C. (1990). *Institutions, institutional change and economic performance.* Cambridge, UK: Cambridge university press.

North, D. C., & Wallis, J. J. (1994). Integrating Institutional Change and Technical Change in Economic History A Transaction Cost Approach. *Journal of Institutional and Theoretical Economics, 150*(4), 609–624.

OECD. (1991). *The nature of innovation and the evolution of the productive system. Technology and productivity - the challenge for economic policy.* Paris.

OECD. (2011). *OECD Guidelines for Multinational Enterprises.* OECD Publishing. Retrieved from http://www.oecd-ilibrary.org/governance/oecd-guidelines-for-multi national-enterprises_9789264115415-en

OECD. (2014). *OECD Science, Technology and Industry Outlook 2014.* OECD Publishing. Retrieved from Industry Outlook 2014, OECD Publishing. http://dx.doi.org/ 10.1787/sti_outlook-2014-en

Oke, A., Burke, G., & Myers, A. (2007). Innovation types and performance in growing UK SMEs. *International Journal of Operations & Production Management, 27*(7), 735–753.

Osawa, J., & Mozur, P. (2014, January 16). The Rise of China's Innovation Machine. *Wall Street Journal*. Retrieved from http://www.wsj.com/articles/SB1000142405270 23038197045793205442313 96168

Palepu, K. G., Khanna, T., & Vargas, I. (2006). Haier: Taking a Chinese company global. *Harvard Business School*, (Case 706-401).

Park, S. H., & Luo, Y. (2001). Guanxi and organizational dynamics: Organizational networking in Chinese firms. *Strategic Management Journal, 22*(5), 455–477.

Peng, M. W. (2002). Towards an institution-based view of business strategy. *Asia Pacific Journal of Management, 19*(2-3), 251–267.

Peng, M. W., & Luo, Y. (2000). Managerial ties and firm performance in a transition economy: The nature of a micro-macro link. *Academy of Management Journal, 43*(3), 486–501.

Peng, M. W., Sun, S. L., Pinkham, B., & Chen, H. (2009). The Institution-Based View as a Third Leg for a Strategy Tripod. *The Academy of Management Perspectives, 23*(3), 63–81.

Peng, M. W., Wang, D. Y. L., & Jiang, Y. (2008). An institution-based view of international business strategy: a focus on emerging economies. *Journal of International Business Studies, 39*(5), 920–936.

Pettigrew, A. M. (1990). Longitudinal field research on change: theory and practice. *Organization Science, 1*(3), 267–292.

Pharmaceutical Manufacturers Association USA. (2015). Retrieved January 15, 2015, from http://www.phrma.org/

Porter, M. E. (1980). *Competitive strategy: Techniques for analyzing industries and companies*. New York: The Free Press.

Porter, M. E. (1990). *The competitive advantage of nations: creating and sustaining superior performance*. New York: The Free Press.

Prahalad, C. K. K., Di Benedetto, A., & Nakata, C. (2012). Bottom of the pyramid as a source of breakthrough innovations. *Journal of Product Innovation Management, 29*(1), 6–12.

Prud'homme, D. (2012). *Dulling the Cutting Edge: How Patent-Related Policies and Practices Hamper Innovation in China*. European Union Chamber of Commerce in China.

Prud'homme, D. (2015). Forecasting threats and opportunities for foreign innovators in China's strategic emerging industries: a policy-based analysis. *Thunderbird International Business Review, forthcomin*.

Ramo, J. C. (2004). *The Beijing Consensus*. London: Foreign Policy Centre .

Rauch, J. E., & Trindade, V. (2002). Ethnic Chinese networks in international trade. *Review of Economics and Statistics, 84*(1), 116–130.

Ray, P. K., & Ray, S. (2010). Resource-constrained innovation for emerging economies: The case of the Indian telecommunications industry. *IEEE Transactions on Engineering Management, 57*(1), 144–156.

Robertson, R. (1995). Glocalization: Time-space and homogeneity-heterogeneity. In M. Featherstone, S. Lash, & R. Robertson (Eds.), *Global modernities* (1st ed., pp. 25–44). London: Sage.

Rotter, E. (2013). *Deutsche Automobilindustrie in China weiter auf Wachstumskurs*.

Rowley, J., Baregheh, A., & Sambrook, S. (2011). Towards an innovation-type mapping tool. *Management Decision, 49*(1), 73–86. Retrieved from http://www.emeraldin sight.com/10.1108/00251741111094446

Roy, W. G. (1999). *Socializing capital: The Rise of the Large Industrial Corporation in America*. New Jersey: Princton University Press.

Sachs, J. D. (2003). *Institutions don't rule: direct effects of geography on per capita income*. National Bureau of Economic Research.

Sandberg, L. G. (1979). The case of the impoverished sophisticate: human capital and Swedish economic growth before World War I. *The Journal of Economic History, 39*(01), 225–241.

Savage, C., & Barker, T. C. (2012). *Economic History of Transport in Britain*. Routledge.

Savitt, R. (1980). Why Study Marketing History? *Journal of Marketing, 44*(4), 52–58.

Schumpeter. (2013). *Capitalism, Socialism and Democracy*. Routledge.

Schumpeter, J. (1934). *The theory of economic development: an inquiry into profits, capital, credit, interest, and the business cycle*. London: Oxford University Press.

Schumpeter, J. A. (1939). *Business Cycles: A Theoretical, Historical, and Statistical Analysis of the Capitalist Process* (Vol. 1). New York and London: McGraw-Hill.

Schumpeter, J. A. (1942). Capitalism and the Process of Creative Destruction. In *Monopoly Power and Economic Performance* (pp. 19–38).

Scott, R. (1995). *Institutions and Organizations. Ideas, Interests and Identities*. Sage.

Scott, W. R. (1987). The adolescence of institutional theory. *Administrative Science Quarterly, 32*(4), 493–511.

Shin, J.-S. (2013). *The Economics of the Latecomers: Catching-up, technology transfer and institutions in Germany, Japan and South Korea*. Routledge.

Siemens AG Corporate Website. (2015). Retrieved January 15, 2015, from http://www.siemens.com/entry/cc/en/

Siemens R&D in China. (2015). Retrieved January 15, 2015, from http://w1.siemens.com.cn/en/about_us/research.asp

Signorini, P., Wiesemes, R., & Murphy, R. (2009). Developing alternative frameworks for exploring intercultural learning: a critique of Hofstede's cultural difference model. *Teaching in Higher Education, 14*(3), 253–264.

SIPO. (2014). Chinese patent database. Retrieved October 6, 2014, from http://www.sipo.cn

Skorna, A., & Widenmayer, B. (2010, April). Komplexe IT-Projekte systematisch strukturieren. *Wissenschaftsmanagement*, 37–41.

Sobel, R. S. (2008). Testing Baumol: Institutional quality and the productivity of entrepreneurship. *Journal of Business Venturing, 23*(6), 641–655.

State Council, S. Decision on Accelerating Development of Strategic Emerging Industries (2010).

Suchman, M. C. (1995). Managing legitimacy: Strategic and institutional approaches. *Academy of Management Review, 20*(3), 571–610.

Sun, Y., Von Zedtwitz, M., & Fred Simon, D. (2007). Globalization of R&D and China: an introduction. *Asia Pacific Business Review, 13*(3), 311–319.

Tan, J., & Tan, D. (2005). Environment–strategy co-evolution and co-alignment: a staged model of Chinese SOEs under transition. *Strategic Management Journal, 26*(2), 141–157.

Taylor, M. Z., & Wilson, S. (2012). Does culture still matter?: The effects of individualism on national innovation rates. *Journal of Business Venturing, 27*(2), 234–247.

Tellis, G. J. (2006). Disruptive Technology or Visionary Leadership? *Journal of Product Innovation Management, 23*(1), 34–38.

Thorstein, V. (1915). *Imperial Germany and the Industrial Revolution.* New York: The Macmillan Company.

Trompf, G. W. (1979). *The idea of historical recurrence in Western thought: from antiquity to the Reformation* (Vol. 1). University of California Press.

Tunzelmann, G. N. von. (1978). *Steam power and British industrialization to 1860.* Oxford: Clarendon Press.

University of Southern California. (2014). Historical Research Method. Retrieved December 15, 2014, from http://ecu.au.libguides.com/historical

USCBC. (2013). *China's Strategic Emerging Industries: Policy, Implementation, Challenges, & Recommendations. US-China Business Council.*

USCBC, 2014. (2014). USCBC 2014 China Business Environment Survey Results: Growth Continues Amidst Rising Competition , Policy Uncertainty USCBC 2014 China Business Environment Survey Results.

Van Hoorn, A., & Maseland, R. (2013). Does a Protestant work ethic exist? Evidence from the well-being effect of unemployment. *Journal of Economic Behavior & Organization, 91*, 1–12.

Van Someren, T. C. R., & van Someren-Wang, S. (2013). *Innovative China: Innovation Race Between East and West.* Berlin Heidelberg: Springer. Retrieved from http://www.dandelon.com/servlet/download/attachments/dandelon/ids/DE0106F0EECCB67473122C1257B83004B5F91.pdf

Vernon, R. (1966). International investment and international trade in the product cycle. *The Quarterly Journal of Economics, 80*(2), 190–207.

Von Zedtwitz, M. (2004). Managing foreign R&D laboratories in China. *R&D Management, 34*(4), 439–452.

Von Zedtwitz, M., Corsi, S., Søberg, P. V., & Frega, R. (2015). A Typology of Reverse Innovation. *Journal of Product Innovation Management, 32*(1), 12–28. Retrieved from http://dx.doi.org/10.1111/jpim.12181

Walsh, V. (1984). Invention and Innovation in the Chemical Industry: Demand-pull or Discovery-push? *Research Policy, 13*(4), 211–234.

Wang, H., & Kimble, C. (2010). Betting on Chinese electric cars? Analysing BYD's capacity for innovation. *International Journal of Automotive Technology and Management, 10*(1), 77.

Wang, A; Liao, W. H., & McKinsey & Company. (2012). Bigger, better, broader: A perspective on China's auto market in 2020. Retrieved September 15, 2014, from http://www.mckinsey.com/client_service/automotive_and_assembly/latest_thinking

Weber, M. (1904). *Die protestantische Ethik und der Geist des Kapitalismus.* JCB Mohr (Paul Siebeck).

Whitehead, A. (1925). 1967. Science and the Modern World. Macmillan. New York.

Why History Matters to Managers. (1986). *Harvard Business Review*, (January).

Widenmayer, B. L. P. (2012). *Reverse Innovation: Insights from Western Medical Equipment Manufacturers in China* (1st ed.).

Wolcott, S. (2010). An examination of the supply of financial credit to entrepreneurs in Colonial India. In D. S. Landes, J. Mokyr, & W. J. Baumol (Eds.), *The Invention of Enterprise: Entrepreneurship from Ancient Mesopotamia to Modern Times* (pp. 443–468). Princeton and Oxford: Princeton University Press.

Worldbank. (2010, January 19). State-owned enterprises in China: How big are they? Retrieved October 15, 2014, from http://blogs.worldbank.org/eastasiapacific/state-owned-enterprises-in-china-how-big-are-they

Worldbank Glossary. (2014). Retrieved August 15, 2014, from http://www.worldbank.org/depweb/english/beyond/global/glossary.html

Yang, J., Liu, H., Gao, S., & Li, Y. (2010). Technological innovation of firms in China: Past, present, and future. *Asia Pacific Journal of Management, 29*(3), 819–840. Retrieved from http://link.springer.com/10.1007/s10490-010-9243-3

Yin, R. K. (1989). Case Study Research: Design And Methods (Applied Social Research Methods) Author: Robert K. Yin, Publisher: Sage Publicat. Sage Publications, Inc.

Yin, R. K. (2014). *Case study research: Design and methods.* Sage publications.

Yip, G., & McKern, B. (2014). Innovation in emerging markets – the case of China. *International Journal of Emerging Markets, 9*(1), 2–10. Retrieved from http://www.emeraldinsight.com/doi/abs/10.1108/IJoEM-11-2013-0182

Yu, D., & Hang, C. C. (2010). A Reflective Review of Disruptive Innovation Theory. *International Journal of Management Reviews, 12*(4), 435–452.

Zeschky, M., Widenmayer, B., & Gassmann, O. (2011). Frugal innovation in emerging markets. *Research-Technology Management, 54*(4), 38–45.

Zeschky, M., Widenmayer, B., & Gassmann, O. (2014). Organising for reverse innovation in Western MNCs: the role of frugal product innovation capabilities. *International Journal of Technology Management, 64,* 255.

Zhang, R. (2009). The Chief Executive. Retrieved August 15, 2014, from http://www.forbes.com/2009/09/25/haier-zhang-ruimin-china-leadership-ruimin.html

Zheng Zhou, K. (2006). Innovation, imitation, and new product performance: The case of China. *Industrial Marketing Management, 35*(3), 394–402.

Zhou, K. Z., Yim, C. K., & Tse, D. K. (2005). The effects of strategic orientations on technology-and market-based breakthrough innovations. *Journal of Marketing, 69*(2), 42–60.

Zhu, Y., Wittmann, X., & Peng, M. W. (2012). Institution-based barriers to innovation in SMEs in China. *Asia Pacific Journal of Management, 29*(4), 1131–1142.

Zucker, L. G. (1987). Institutional theories of organization. *Annual Review of Sociology, 13,* 443–464.

Appendix

A. Questionnaire for case study interviews conducted in China

PhD Thesis, Joachim Thraen   Survey: *Face-to-face interviews (China)*

Innovation in China's Strategic Emerging Industries
Integrating the Historical Perspective

<u>*Company:*</u>   *Date:*

Interviewee name:
Position:
Background:
Location:

Interviewer:
Joachim Jan Thraen
Research Associate and PhD Candidate, Asia Research Centre, University of St. Gallen
http://www.arc.unisg.ch

Relevant work experience:
- Associate Consultant, Bain & Company; innovation strategy projects, industrial goods clients (CH/SWE/US/MEX)
- Consultant, International Trade Centre (WTO/UN) Geneva; research project on state-created non-tariff-measures imposing barriers to export companies in developing countries

Joachim Jan Thraen
Research Associate
PhD Candidate, HSG St. Gallen

1. **Please provide your perspective of your firm's position and recent activities in the China market, related to R&D and innovation.**

2. **Which SEI is your firm operating in and how to you view China's more recent SEI policies to increase levels of indigenous innovation in China?**

3. **What do you perceive as the most important challenges in doing R&D and innovation activities in China?**

   - Provide examples if possible such as:
     o Protection of intellectual property rights
     o Government policies (e.g. lack of transparency)
     o Human resources
     o Others:_____:

4. **What do you perceive as the most important challenges in doing R&D and innovation activities in China?**

5. **How do you evaluate China's innovation environment in the recent past, present and what is your personal opinion about its future?**

6. **In your opinion, which role does the Chinese government (central and local) have in promoting innovation in China?**

7. **Which impact will government policies to promote indigenous innovation (e.g. in SEIs) have on the future competitiveness of**

   - **Chinese** MNCs operating in SEI
   - **Foreign** MNCs operating in SEI

8. **In your opinion, what are the major economic, institutional-political and sociocultural factors that characterize the current innovation system in China?**

   - Economic factors
   - Institutional/political factors
   - Sociocultural factors

9. **How would you describe your (global) R&D strategy, in relation to China?**

10. **Please share some of your best practices for managing R&D and innovation in China**

11. Further remarks?

12. Available for further questions?

Joachim Jan Thraen
Research Associate
PhD Candidate, HSG St. Gallen

B. Case study interview schedule

Company	SEI/Industry	HQ location	Interview Type	Position	Interview location	Interview date
Bayer Material Science	New Materials	Leverkusen (Germany)	Personal interview	Senior R&D Manager	Shanghai	22-Oct-2014
Bayer Material Science	New Materials	Leverkusen (Germany)	Personal interview	Vice President Application Development and R&D	Shanghai	18-Nov-2014
Bayer Material Science	New Materials	Leverkusen (Germany)	Personal interview	Manager government affairs	Shanghai	18-Nov-2014
Bayer Material Science	New Materials	Leverkusen (Germany)	Personal interview	Senior R&D Manager	Shanghai	18-Nov-2014
Bayer Material Science	New Materials	Leverkusen (Germany)	Personal interview	Vice President Application Development and R&D	Shanghai	24-Oct-2014
BYD-Daimler	Automotive	Shenzhen	Phone Interview	Director, New Electric Vehicles	St. Gallen	3-Dec-2014
BYD-Daimler	Automotive	Stuttgart (Germany)	Personal interview	Director - Safety, Comfort, Regulatory Affairs & Intellectual Property	Beijing	10-Sep-2014
BYD-Daimler	Automotive	Stuttgart (Germany)	Personal interview	Senior Manager, Design & Innovation	Beijing	18-Nov-2014
BYD-Daimler	Automotive	Stuttgart (Germany)	Personal interview	Manager, Design & Innovation	Beijing	19-Nov-2014
Haier	Industrial goods	Qingdao (China)	Personal interview	Manager, Open Innovation	Shanghai	24-Oct-2014
Haier	Industrial goods	Qingdao (China)	Phone Interview	Senior R&D Manager	Shanghai	14-Nov-2014
Haier	Industrial goods	Qingdao (China)	Phone Interview	Manager, Innovation Cooperation	St. Gallen	4-Dec-2014
Siemens	Machinery	Munich (Germany)	Personal interview	Head of Research	Beijing	9-Sep-2014
Siemens	Machinery	Munich (Germany)	Phone Interview	Senior Innovation Manager	Phone	16-Oct-2014
Siemens	Machinery	Munich (Germany)	Personal interview	Head of Research	Beijing	17-Nov-2014
Siemens	Machinery	Munich (Germany)	Personal interview	Senior Innovation Manager	Beijing	17-Nov-2014
Siemens	Machinery	Munich (Germany)	Personal interview	Innovation Manager	Beijing	17-Nov-2014